M336
Mathematics and Computing: a third-level course

GROUPS & GEOMETRY

UNIT GE4
WALLPAPER PATTERNS

Prepared for the course team by
David Asche & Fred Holroyd

The Open University

This text forms part of an Open University third-level course.
The main printed materials for this course are as follows.

Block 1
Unit IB1   Tilings
Unit IB2   Groups: properties and examples
Unit IB3   Frieze patterns
Unit IB4   Groups: axioms and their consequences

Block 2
Unit GR1   Properties of the integers
Unit GR2   Abelian and cyclic groups
Unit GE1   Counting with groups
Unit GE2   Periodic and transitive tilings

Block 3
Unit GR3   Decomposition of Abelian groups
Unit GR4   Finite groups 1
Unit GE3   Two-dimensional lattices
Unit GE4   Wallpaper patterns

Block 4
Unit GR5   Sylow's theorems
Unit GR6   Finite groups 2
Unit GE5   Groups and solids in three dimensions
Unit GE6   Three-dimensional lattices and polyhedra

The course was produced by the following team:

Andrew Adamyk (BBC Producer)
David Asche (Author, Software and Video)
Jenny Chalmers (Publishing Editor)
Bob Coates (Author)
Sarah Crompton (Graphic Designer)
David Crowe (Author and Video)
Margaret Crowe (Course Manager)
Alison George (Graphic Artist)
Derek Goldrei (Groups Exercises and Assessment)
Fred Holroyd (Chair, Author, Video and Academic Editor)
Jack Koumi (BBC Producer)
Tim Lister (Geometry Exercises and Assessment)
Roger Lowry (Publishing Editor)
Bob Margolis (Author)
Roy Nelson (Author and Video)
Joe Rooney (Author and Video)
Peter Strain-Clark (Author and Video)
Pip Surgey (BBC Producer)

With valuable assistance from:

Maths Faculty Course Materials Production Unit
Christine Bestavachvili (Video Presenter)
Ian Brodie (Reader)
Andrew Brown (Reader)
Judith Daniels (Video Presenter)
Kathleen Gilmartin (Video Presenter)
Liz Scott (Reader)
Heidi Wilson (Reader)
Robin Wilson (Reader)

The external assessor was:
Norman Biggs (Professor of Mathematics, LSE)

The Open University, Walton Hall, Milton Keynes, MK7 6AA.

First published 1994. Reprinted 1997, 2002, 2005, 2009.

Copyright © 1994 The Open University

All rights reserved. No part of this publication may be reproduced, stored in a retrieval system or transmitted in any form or by any means, without written permission from the publisher or a licence from the Copyright Licensing Agency Limited. Details of such licences (for reprographic reproduction) may be obtained from the Copyright Licensing Agency Ltd of 90 Tottenham Court Road, London, W1P 9HE.

Edited, designed and typeset by the Open University using the Open University TeX System.

Printed in Malta by Gutenberg Press Limited.

ISBN 0 7492 2172 0

This text forms part of an Open University Third Level Course. If you would like a copy of *Studying with the Open University*, please write to the Central Enquiry Service, PO Box 200, The Open University, Walton Hall, Milton Keynes, MK7 6YZ. If you have not already enrolled on the Course and would like to buy this or other Open University material, please write to Open University Educational Enterprises Ltd, 12 Cofferidge Close, Stony Stratford, Milton Keynes, MK11 1BY, United Kingdom.

CONTENTS

Study guide	4
Introduction	5
1 Wallpaper patterns	**6**
1.1 Basic ideas	6
1.2 The wallpaper groups	9
2 Classification of patterns	**13**
2.1 Types of pattern	13
2.2 The point group of a wallpaper pattern	16
2.3 Direct and indirect symmetries	20
3 Parallelogram and rectangular patterns	**25**
3.1 Parallelogram patterns: $p2$ and $p1$	25
3.2 Pattern type $p2mm$	27
3.3 Pattern type $p2mg$	28
3.4 Pattern type $p2gg$	30
3.5 Pattern type pm	32
3.6 Pattern type pg	33
4 Rhombic and square patterns	**34**
4.1 Pattern type $c2mm$	35
4.2 Pattern type cm	36
4.3 Pattern type $p4mm$	37
4.4 Pattern type $p4gm$	38
4.5 Pattern type $p4$	40
5 Hexagonal patterns	**41**
5.1 Pattern type $p6mm$	41
5.2 Pattern type $p6$	42
5.3 Pattern type $p3m1$	43
5.4 Pattern type $p31m$	44
5.5 Pattern type $p3$	45
6 An algorithm (audio-tape section)	**46**
Solutions to the exercises	55
Objectives	63
Index	64

STUDY GUIDE

Sections 1 and 2, which set up all the mathematical results and techniques necessary for an analysis of the symmetry groups of wallpaper patterns, are probably the heaviest in terms of the study time which you should plan to devote: they need to be taken slowly. Sections 3 to 5 apply these techniques and go through all the possibilities. If you find yourself short of time, you are advised to read through these sections fairly quickly.

The crux of the unit is the algorithm in Section 6 which is associated with an audio programme. For this section you will also need the *Unit GE4* Tape Frame Overlays from the *Geometry Envelope*.

You are advised to have the Isometry Toolkit card to hand when studying Sections 1 and 2 of the unit.

The video programme, VC3A *Lattices and Wallpaper Patterns*, should be viewed before you study Section 6.

INTRODUCTION

In this unit, we shall be studying patterns which are often called wallpaper patterns. The common feature of these patterns is that they all possess the translational symmetries of some lattice. They may have other symmetries as well and these additional symmetries can be rotations, reflections or glide reflections. Although we refer to them as wallpaper patterns, we are not claiming that these patterns are exclusive to wallpaper. It is probably easier to discover examples of these patterns by looking at floor tiles or carpets. They occur frequently in the artistic and decorative works which have been created over the centuries by many different civilizations. At the Alhambra in Granada, Spain, there are numerous examples of these periodic designs among the tile patterns created by the Moors. The Dutch artist M.C. Escher was inspired by these patterns and used their symmetries to create his fascinating collection of interlocking designs.

The M336 cover designs are based on patterns from the Alhambra.

We know, from *Unit GE3*, that there are exactly five different types of plane lattice. From each of these, we can produce a variety of wallpaper patterns. In this unit, we shall find that there are precisely seventeen different kinds of wallpaper pattern, the symmetries of each of which form a subgroup of some lattice group. There are, of course, any number of designs which we could use to create patterns, so we need to decide when two wallpaper patterns are to be regarded as equivalent.

In Section 1, we discuss the connection between wallpaper patterns and lattices. We note that a plane lattice is itself a simple kind of wallpaper pattern, and that the five types of plane lattice therefore give us five types of wallpaper patterns, with distinct symmetry groups. We introduce a notation for these groups.

In Section 2, we examine more closely what we mean by the 'type' of a pattern. Then we introduce what is called the *point group* of a wallpaper pattern, and show that this must be one of the ten groups $C_1, C_2, C_3, C_4, C_6,$ D_1, D_2, D_3, D_4, D_6. We then study the direct and indirect symmetries of a wallpaper pattern, analysing the way in which reflection axes and glide axes can occur. These considerations allow us to prove that there are indeed just seventeen types of wallpaper pattern.

In Sections 3–5, we examine the symmetries and study the symmetry group for each type of pattern. We also give examples of each type of wallpaper pattern. These examples have been chosen more for clarity than for artistic merit. It is easier to understand the symmetries when the design is fairly simple. We leave it to you to make up your own exotic patterns.

In the last section, we start by providing a method for identifying a given wallpaper pattern as one of the seventeen types. After this, we provide a summary of the facts that we have discovered, listing all the groups of symmetries of the various patterns and giving further details.

If you have access to a computer which runs Microsoft *Windows*, you can use the software provided with the course to create numerous wallpaper patterns.

1 WALLPAPER PATTERNS

1.1 Basic ideas

Before we start our investigations, we must make clear what is meant by a wallpaper pattern. We give the following definition.

Definition 1.1 Wallpaper pattern

A **wallpaper pattern** is a subset of the plane whose translational symmetries are those of some lattice.

As in earlier units, we shall use $\mathbb{R}^2$ to represent the plane. Suppose that a wallpaper pattern W has the same translations as a lattice L with basis $\{\mathbf{a}, \mathbf{b}\}$. If $\mathbf{p}$ is a point of W, then the points $\mathbf{p} + n\mathbf{a} + m\mathbf{b}$, where n and m are integers, will all be points of W. In our illustrations, we show the points of the pattern in black and the remaining points of the plane in white. Note that a symmetry of a wallpaper pattern is any isometry that maps black points to black points and white points to white points. It will be clear that the pattern W is fully determined by the points that lie in some basic parallelogram of the lattice L since we can obtain all the other points of the pattern by applying the various translations of L.

Thus, W in this unit refers to the black points. W does not mean white, as it did in Unit GE1!

This may all seem rather colourless! Most wallpaper patterns actually use several colours, and the symmetry group of a multicoloured pattern should presumably be defined as the group of isometries that map every point of $\mathbb{R}^2$ to a point *of the same colour*. This is true enough; but in fact the range of possibilities for the symmetry group of a multicoloured pattern is the same as for a black and white pattern. Thus, we can make a complete analysis of the symmetries of wallpaper patterns by restricting our attention to black and white patterns and analysing the symmetry groups of the inked subsets of $\mathbb{R}^2$.

We shall not give a mathematical proof of this fact; but you should be able to see that it is a reasonable result, since one can convert a multicoloured pattern into a black and white one in which each colour is represented by a different style of shading.

When we studied lattices in *Unit GE3*, we always chose the origin O to be one of the lattice points. In the case of wallpaper patterns, there is no such natural choice for the origin. Any point of the plane may be chosen but, in many cases, it will be convenient to take the origin to be the centre for some rotation, or a point on some reflection or glide axis. Although a wallpaper pattern W will have the translational symmetries of some lattice L, you should note that the lattice points themselves need not be points of the pattern.

In our investigations of wallpaper patterns, we shall need to refer to a lattice that has the same translations as the pattern. It will be useful to have a name for such a lattice, so we make the following definition.

Definition 1.2 Associated lattice

A plane lattice L whose translations are the same as those of a wallpaper pattern W is said to be **associated** with W.

Once we have chosen the origin O for the plane containing our wallpaper pattern, there will be exactly one lattice associated with the pattern that has O as one of its lattice points. This is because a lattice is fully determined by a single lattice point and its set of translations. For a given wallpaper pattern W and a point O, we shall call this lattice the **associated lattice based at** O.

We shall need the idea of a *basic parallelogram* of a wallpaper pattern, so we define this now.

> **Definition 1.3 Basic parallelogram**
>
> A **basic parallelogram** of a wallpaper pattern is a basic parallelogram of some associated lattice.

Since we are free to choose the origin O as we please and since we may use any basis for the lattice, there will be any number of basic parallelograms that we could take.

Example 1.1

In Figure 1.1, we show a simple pattern together with various choices for a basic parallelogram.

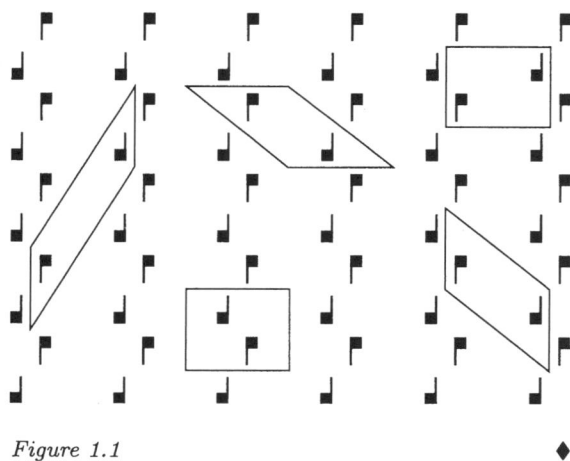

Figure 1.1 ♦

By a **design** for a wallpaper pattern, we mean that portion of a wallpaper pattern which is contained in some basic parallelogram of the pattern. You can see that every part of a design will occur somewhere within any basic parallelogram that you may choose. From any design, we may complete the whole pattern by applying the translations of the lattice to the basic parallelogram.

One way of constructing a wallpaper pattern would be to draw a design on a basic parallelogram of some lattice and then extend it to cover the whole plane using the translations of the lattice. In many cases, you will find that the resulting pattern has fewer symmetries than the lattice. On the other hand, it is possible to introduce symmetries that the original lattice does not possess. This may happen in various ways.

Example 1.2

Take, for example, a rectangular lattice L. On the basic rectangle shown in Figure 1.2, let our design consist of dots at the four vertices and a dot at the centre. The pattern constructed from this design will have more translational symmetries than the lattice L. For example, the translation that sends a vertex of the rectangle to its centre will be a translation of the pattern but not of the lattice L. Thus the lattice L will not be an associated lattice for this pattern. All the lattices associated with this pattern will be rhombic rather than rectangular.

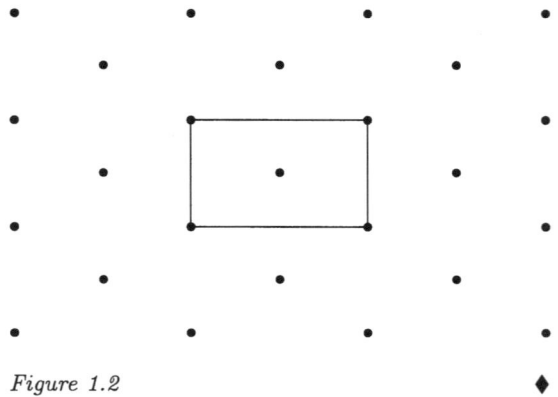

Figure 1.2   ♦

Example 1.3

We can achieve extra symmetries even without changing the associated lattice. Again take a rectangular lattice L. Our basic design consists of the letter R and its mirror image placed as shown in Figure 1.3. When we apply the translations of L to this design, the resulting wallpaper pattern has precisely the translational symmetries of L. This means that L is an associated lattice for the pattern. Nevertheless, the pattern has some symmetries that are not possessed by the lattice L. In the next exercise, we ask you to find one.

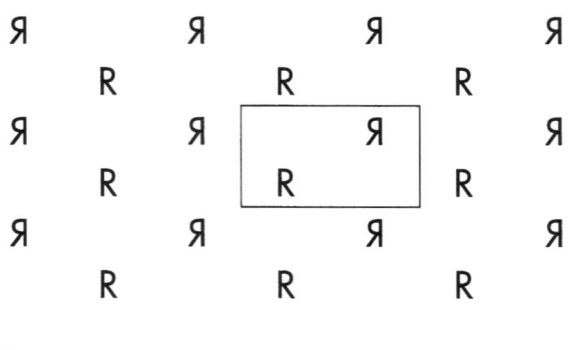

Figure 1.3   ♦

Exercise 1.1

Find a glide reflection symmetry of the pattern W in Figure 1.3 that is not a symmetry of any associated lattice.

Another useful idea is that of a *generating region* for a wallpaper pattern. We shall define this now.

> **Definition 1.4 Generating region**
>
> A **generating region** for a wallpaper pattern W is a smallest region of the plane such that every point in the plane lies in some image of the region under the symmetries of W.

In Figure 1.4, we show part of a wallpaper pattern whose basic parallelogram is a square. The shaded triangle is a generating region of the pattern. To see this, consider the symmetries of the pattern that fix the centre of the basic square. They form the dihedral group D_4 of order 8, which is the symmetry group of a square. Each of these symmetries will map the shaded triangle to another triangle within the square. The eight congruent triangles that we obtain in this way cover the square. Applying all the translations of the pattern to this square enables us to cover the whole plane. It is possible to prove that no smaller region will suffice.

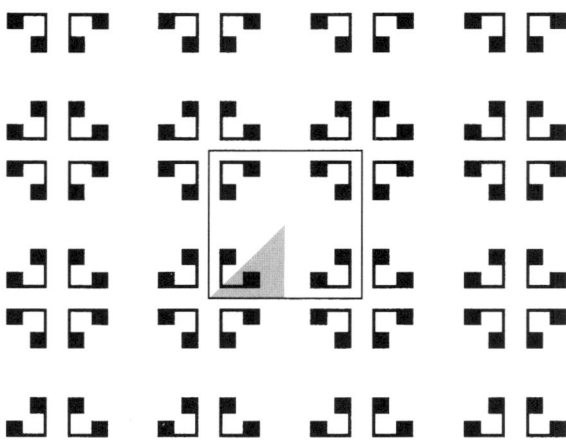

Figure 1.4

Once you have become familiar with the various types of wallpaper patterns and know their symmetries, it is easy to construct wallpaper patterns using a suitable generating region. You will still have to be careful not to introduce additional symmetries within the generating region, otherwise there is the danger that you will end up with a different type of wallpaper pattern.

Exercise 1.2

Find a generating region for the wallpaper pattern in Figure 1.5. (The rectangle drawn in the figure is a basic rectangle.)

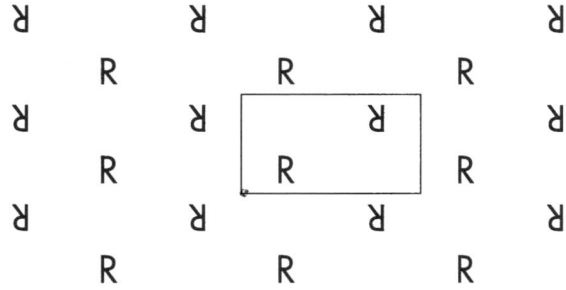

Figure 1.5

Exercise 1.3

Find a generating region for the wallpaper pattern shown in Figure 1.3.

In our study of patterns, we shall make use of these generating regions to help us find the orbits of various points under the action of the appropriate symmetry group. From the definition, we can see that every point in the plane can be mapped into a generating region by a suitable symmetry of the pattern. This means that a generating region will contain a representative from each orbit of points under the action of the symmetry group.

1.2 The wallpaper groups

The symmetry groups of wallpaper patterns are known as *wallpaper groups*. This is an important enough concept to be worth emphasizing.

> *Definition 1.5 Wallpaper group*
>
> A **wallpaper group** is a group of plane isometries that is the symmetry group of some wallpaper pattern.

In the course of this unit, we shall see that there are seventeen distinct types of wallpaper group. You have already seen five of them in *Unit GE3*, under the guise of symmetry groups of plane lattices. This is because the simple pattern consisting of a dot at every point of a plane lattice is an example of a wallpaper pattern.

Figure 1.6 shows the five types of plane lattice.

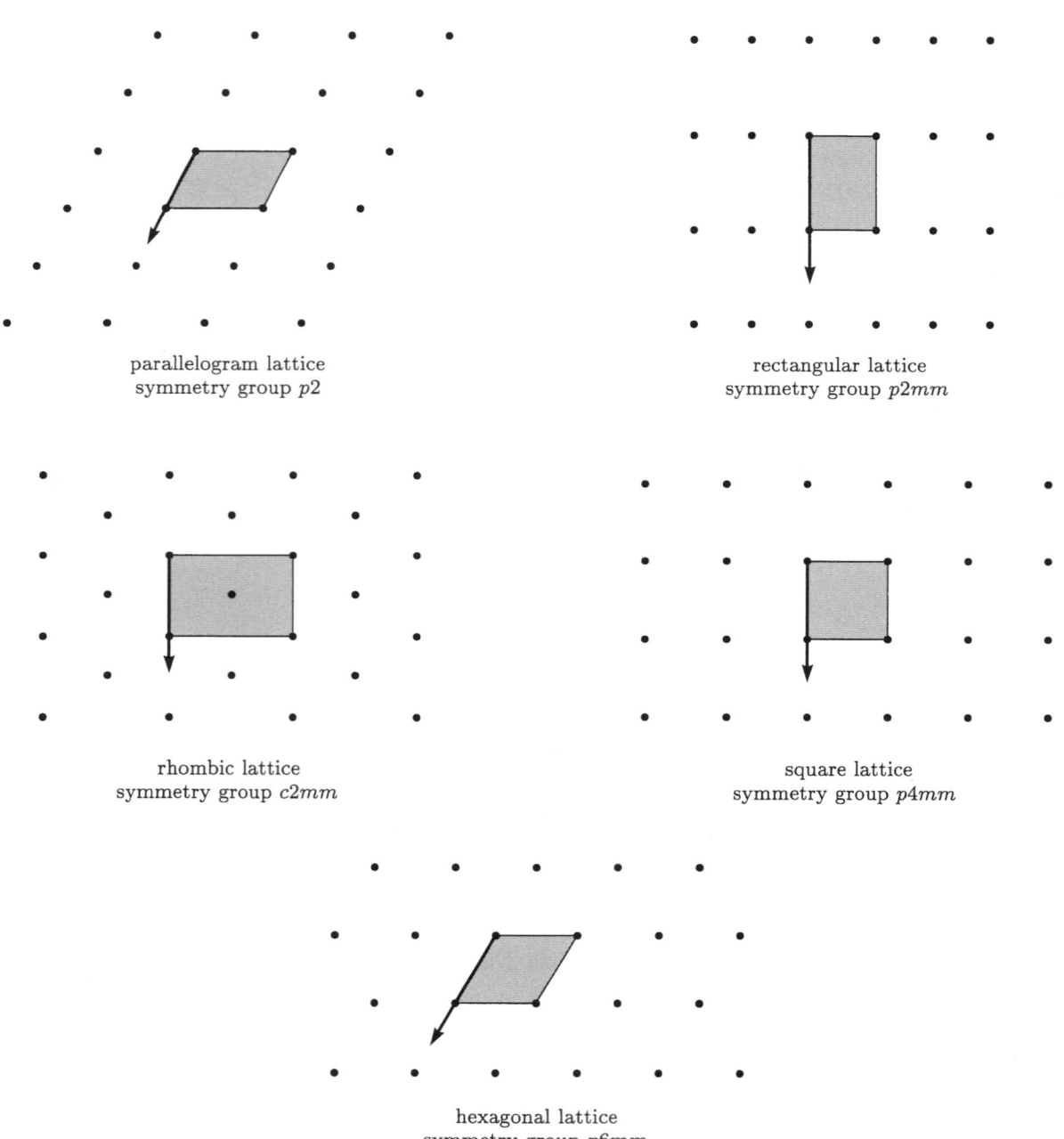

Figure 1.6

Just as with the frieze groups, there is a reasonably well-established system of notation for the wallpaper groups. It is not completely standardized, and there are slight differences between different authors. What we shall do is to describe a notation very close to that used by the late Rolf Schwarzenberger of the University of Warwick; at the end of the unit, we shall compare it with the more abbreviated notation used in the International Tables for X-ray crystallography.

Schwarzenberger, R. (1980) 'N-dimensional crystallography', *Research Notes in Mathematics*, **41**, Pitman Advanced Publishing Program.

Our explanation of the concepts behind the notation is based on an article written in 1978 by Doris Schattschneider.

A 'cell' is defined for each lattice (shown shaded in Figure 1.6). This is a basic parallelogram, called a *primitive cell*, for four of the five lattice types; but for the rhombic lattice the cell is a so-called *centred cell*. This is a rectangle of twice the area of a basic parallelogram, with a lattice point at each corner and another at the centre. A *reference direction* is defined in each case, directed *downwards* along the left-hand edge of the cell.

Schattschneider, D. (1978) 'The plane symmetry groups: their recognition and notation', *American Mathematical Monthly*, June–July, pp. 439–50.

The parallelogram lattice

The notation for the symmetry group of this lattice is

$p2$.

The letter p stands for 'primitive cell', and the number 2 indicates that the maximum order of a rotational symmetry is 2.

The rectangular lattice

The notation for the symmetry group of this lattice is

$p2mm$.

The letter p and the number 2 have the same significance as above. The two occurrences of the letter m refer to the two directions (horizontal and vertical) of the axes of reflection.

It is important for future reference to note that the *first* m refers to the existence of reflection axes *perpendicular to the reference direction*. (This may seem a somewhat arbitrary way to refer to a horizontal reflection; but in fact the notation evolved out of the work of crystallographers on three-dimensional symmetry, and a three-dimensional reflection is through a plane, so the direction *perpendicular* to the plane of reflection is the most convenient one to refer to.)

cf. Alice *through* the looking glass!

The rhombic lattice

The notation for the symmetry group of this lattice is

$c2mm$.

The letter c stands for 'centred cell', and the number 2 again indicates that the maximum order of a rotation symmetry is 2. Once again, the two occurrences of m refer to the two directions of reflection axes, the first one being the horizontal axis because it is perpendicular to the reference direction.

The square lattice

The notation for the symmetry group of the square lattice is

$p4mm$.

The letter p stands for 'primitive cell', as for the parallelogram and rectangular lattices. The number 4 records the presence of rotations of order 4.

Care is needed here in interpreting the two occurrences of the letter m, as there are reflection axes in *four* directions rather than two.

As before, the first occurrence of m refers to the horizontal axis at right angles to the reference direction. Given that such an axis exists, however, we automatically know (from the existence of rotations of order 4) that vertical reflection axes must exist as well. Thus the first letter m refers to *both* these directions. The second letter m refers to the fact that there are *also* reflection axes in the directions that bisect the angles between the directions referred to by the *first m*.

The hexagonal lattice

The notation for the symmetry group of the hexagonal lattice is

$p6mm$.

Exercise 1.4

How many directions of reflection axes are there for the hexagonal lattice? Which of them correspond to the first letter m and which to the second letter m?

The symmetry groups of the five lattice types are particularly important, because *every* wallpaper group is a subgroup of the symmetry group of some lattice.

Exercise 1.5

The wallpaper pattern W in Figure 1.7 has a parallelogram lattice for an associated lattice, but W does *not* possess any rotations of order 2. (All the kites are flying the same way up!) Try to guess the notation for this type of wallpaper pattern.

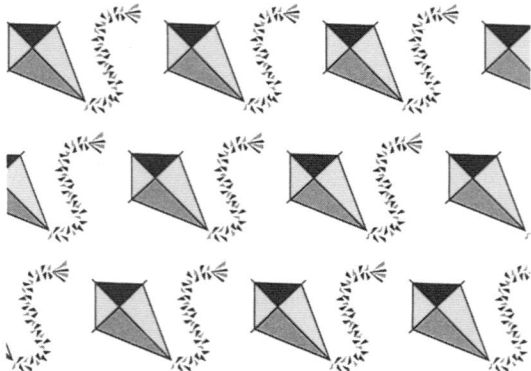

Figure 1.7

2 CLASSIFICATION OF PATTERNS

2.1 Types of pattern

We have, on occasions, talked about various types of wallpaper pattern but we have not yet defined what we mean by the word 'type'. Since there is clearly an unlimited number of possible designs, we need to agree when two patterns are to be regarded as equivalent.

Example 2.1

Let us consider the three designs shown in Figure 2.1.

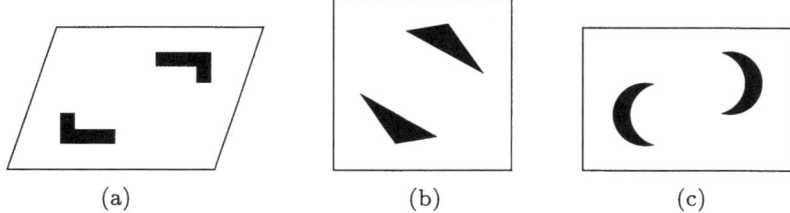

Figure 2.1

The lattices associated with the corresponding wallpaper patterns are clearly different. For design (a), it is a parallelogram lattice, for (b), it is square and for (c), it is rectangular. We might expect that the wallpaper patterns based on these designs should be considered as different types. This is not, however, how we classify wallpaper patterns. The shape of the basic parallelogram is not a deciding feature, nor are the various objects we have drawn. It is the *symmetries of the pattern as a whole* that matter.

Let us see how we can justify saying that the three wallpaper patterns based on the designs above are essentially the same. If you examine each of these patterns, you will find that the only symmetries apart from the translations are rotations of order 2. For each pattern, let us take an associated lattice based at a 2-centre. We can map any one of these associated lattices to any other one by applying a suitable affine transformation ϕ — that is, an affine transformation that maps a basic parallelogram of one to a basic parallelogram of the other. You will notice that, using such a mapping ϕ, the 2-centres of the one pattern will be mapped to the 2-centres of the other pattern, for if r is a rotation of order 2, then so is $\phi r \phi^{-1}$. This means that we can use an affine transformation to set up a one–one correspondence between the symmetries of each of these patterns. It is this fact that makes us say that these patterns are of the same type. ♦

Let us consider another example before giving the formal definition.

Example 2.2

Look at the patterns (a), (b) and (c) shown in Figure 2.2. For each of them, a basic parallelogram contains two copies of the letter F or its mirror image. The lattice associated with each of these patterns is square.

That is, the basic parallelogram is square.

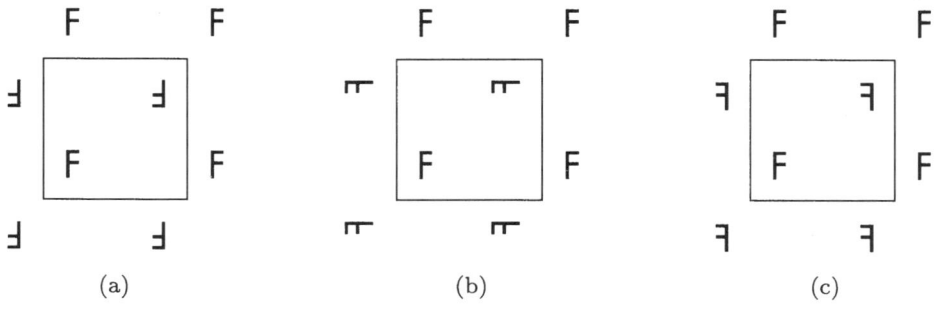

Figure 2.2

Let us examine the symmetries of these patterns. Pattern (a) has rotations of order 2 but no reflections. Pattern (b) has no rotations but has reflections and glide reflections in one of the diagonal directions. Finally, pattern (c) has no rotations or reflections but it does possess glide reflections in some vertical axes. Although the associated lattices are of the same type, the symmetries of these three patterns are quite different. There is certainly no affine transformation which will enable us to set up a one–one correspondence between their symmetries, since such a correspondence must send rotations to rotations, reflections to reflections and glide reflections to glide reflections. We say that these patterns are of different types. ◆

Having seen these two examples, we are now ready for a definition.

> **Definition 2.1 Type of wallpaper pattern**
>
> Suppose W and W' are wallpaper patterns with symmetry groups G and G', respectively. Then W' has the same **type** as W if there is an affine transformation ϕ of the plane such that the mapping $g \mapsto g' = \phi g \phi^{-1}$ is a mapping of G onto G'.

Choosing an origin O for the plane, we may write the translation subgroups of G and G' as $T_2 = \langle t[\mathbf{a}], t[\mathbf{b}] \rangle$ and $T_2' = \langle t[\mathbf{a}'], t[\mathbf{b}'] \rangle$, where

$$t[\mathbf{a}'] = \phi\, t[\mathbf{a}]\, \phi^{-1} \quad \text{and} \quad t[\mathbf{b}'] = \phi\, t[\mathbf{b}]\, \phi^{-1}.$$

Let us put $\phi = t[\mathbf{x}]\, h$, where $t[\mathbf{x}]$ is a translation and h is linear. Then

$$\begin{aligned}
t[\mathbf{a}'] &= \phi\, t[\mathbf{a}]\, \phi^{-1} \\
&= t[\mathbf{x}]\, h\, t[\mathbf{a}]\, h^{-1}\, t[-\mathbf{x}] \\
&= t[\mathbf{x}]\, t[h(\mathbf{a})]\, h h^{-1}\, t[-\mathbf{x}] \\
&= t[\mathbf{x} + h(\mathbf{a}) - \mathbf{x}] \\
&= t[h(\mathbf{a})].
\end{aligned}$$

Similarly, $t[\mathbf{b}'] = t[h(\mathbf{b})]$, so $\mathbf{a}' = h(\mathbf{a})$ and $\mathbf{b}' = h(\mathbf{b})$. Thus,

$$\begin{aligned}
\phi(\mathbf{a}) &= (t[\mathbf{x}]\, h)(\mathbf{a}) \\
&= \mathbf{x} + \mathbf{a}' \\
&= \phi(\mathbf{0}) + \mathbf{a}';
\end{aligned}$$

and, similarly,

$$\phi(\mathbf{b}) = \phi(\mathbf{0}) + \mathbf{b}'.$$

If $L = L(\mathbf{a}, \mathbf{b})$ is the associated lattice based at O for the pattern W, then $\phi(L) = \phi(\mathbf{0}) + L(\mathbf{a}', \mathbf{b}')$ will be the associated lattice based at $\phi(\mathbf{0})$ for the pattern W'.

We shall leave it as an exercise to show that the definition above gives us an equivalence relation.

Exercise 2.1

Show that 'type' defines an equivalence relation.

Let us examine more closely the mapping from G to G' that was specified in Definition 2.1. The following result shows that the mapping has important properties. We do not refer specifically to wallpaper patterns since the result holds for more general figures in the plane.

> **Theorem 2.1**
>
> Let G and G' be the symmetry groups of two figures in the plane and let ϕ be an affine transformation such that the mapping $g \mapsto g' = \phi g \phi^{-1}$ is from G onto G'. Then this mapping is an isomorphism for which symmetries of the same geometric type correspond.

Proof

We first show that the mapping $g \mapsto g'$ is one–one. If $\phi g_1 \phi^{-1} = \phi g_2 \phi^{-1}$, then, applying ϕ^{-1} on the left and ϕ on the right of these expressions, we see that $g_1 = g_2$. Hence the mapping is one–one.

Suppose that g_1 and g_2 belong to G. The image of $g_1 g_2$ will then be $\phi g_1 g_2 \phi^{-1} = (\phi g_1 \phi^{-1})(\phi g_2 \phi^{-1})$, which is the composite of the images of g_1 and g_2. Hence the mapping is an isomorphism from G onto G'.

We now need to check that symmetries of the same geometric type correspond. There are, as we know from Section 5 of *Unit IB1*, six geometric types. They are the identity, the non-trivial translations, the rotations of order 2, the other non-trivial rotations, the reflections and the glide reflections. These are distinguishable by the points and lines that they fix. Now we can easily see that an element g in G will fix a point **c** if and only if the corresponding element $g' = \phi g \phi^{-1}$ fixes the point $\phi(\mathbf{c})$. Let us consider lines: this is where the affine properties of ϕ are needed. If l is a line, then $\phi(l)$ is also a line and if l and m are parallel lines, then $\phi(l)$ and $\phi(m)$ will be parallel. A symmetry g in G will fix a line l if and only if its image $g' = \phi g \phi^{-1}$ fixes the line $\phi(l)$. It is now easy to verify that the mapping $g \mapsto g'$ preserves the geometric type of the symmetry. This completes our proof. ∎

In view of Theorem 2.1, there is no need in practice to look for the existence or non-existence of an affine transformation when trying to decide whether two wallpaper patterns are of the same type. What we shall actually do is to look for the maximum order of a rotation symmetry, and for reflection and glide axes. If these are the same, and their geometric arrangements correspond, then we know that the patterns are of the same type.

When the figures are wallpaper patterns W and W', the affine transformation ϕ will map an associated lattice L for W to an associated lattice $\phi(L)$ for W'. Note that these lattices L and $\phi(L)$ are *not* necessarily of the same type. In Figure 2.1, the patterns based on designs (a), (b) and (c) are of the same type but their associated lattices are different. It can be shown, by applying a suitable affine transformation, that all the various types of wallpaper pattern can be displayed such that their associated lattices are either square or hexagonal.

Exercise 2.2

The wallpaper pattern W in Figure 2.3 has a square associated lattice. Could there be an affine transformation ϕ of the plane, mapping W to a pattern W' of the same type but with an associated lattice of a different type? If so, what is the alternative lattice type?

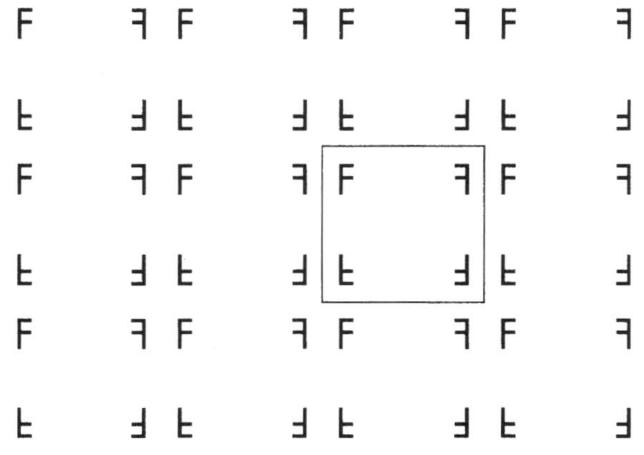

Figure 2.3

Before we can actually embark on the classification of all seventeen wallpaper pattern types, there is further work that we need to do concerning the groups themselves.

2.2 The point group of a wallpaper pattern

In Section 2 of *Unit GE3* we considered the decomposition of a symmetry f of a lattice L. In particular, we saw that we can express f in terms of its translation and linear parts:

$$f = t\,r[\theta] \quad \text{or} \quad t\,q[\theta],$$

and, in this case, t and $r[\theta]$ (or t and $q[\theta]$) are themselves symmetries of L.

What about wallpaper patterns? The situation here turns out to be more complicated. For a start, the decomposition of a symmetry into translation and linear parts depends on our choice of origin, and we have seen that there is no natural or automatic choice of origin in the case of a wallpaper pattern.

Example 2.3

In Figure 2.4, we have repeated the wallpaper pattern W in Figure 1.4, and have chosen a coordinate system whose origin O is *not* at a rotation centre.

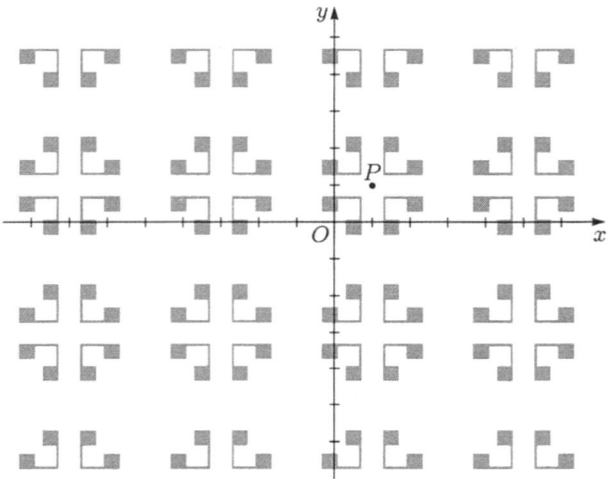

Figure 2.4

Let f be rotation through $\pi/2$ about the point P, which is at a rotation centre of order 4. Then the linear part of f is $r[\pi/2]$, which is not a symmetry of W as O is not a rotation centre. It follows that the translation part of f cannot be a symmetry of W either. ♦

Exercise 2.3

Assuming that P is at the point (1,1), and that the translations of W are generated by $t[(4,0)]$ and $t[(0,4)]$, find the translation part of f and confirm that it is not a symmetry of W.

Thus the translation and linear parts of a wallpaper symmetry are not necessarily themselves symmetries of the pattern. However, the *linear* part is always a symmetry of *the associated lattice based at O*. We shall now prove this.

Theorem 2.2

Let W be a wallpaper pattern, L an associated lattice based at O, and f a symmetry of W. Then the linear part of f is a symmetry of L.

Proof

Express f in standard form, as

$$f = t\,r[\theta] \quad \text{or} \quad f = t\,q[\theta],$$

so that t is the translation part of f, and $r[\theta]$ or $q[\theta]$ the linear part.

We shall prove that each point of L is mapped by $r[\theta]$ or $q[\theta]$ to some other point of L.

Let $\mathbf{x}$ be any point of L, so that the translation $t[\mathbf{x}]$ belongs to the translation group $\Delta(L)$. Since L is associated with W, it follows that $t[\mathbf{x}]$ is a symmetry of W:

$$t[\mathbf{x}] \in \Delta(W).$$

The conjugate of $t[\mathbf{x}]$ by f is the symmetry

$$f\,t[\mathbf{x}]\,f^{-1},$$

and is another translational symmetry of W:

$$f\,t[\mathbf{x}]\,f^{-1} = t[\mathbf{y}],$$

where $\mathbf{y} = r[\theta](\mathbf{x})$ or $\mathbf{y} = q[\theta](\mathbf{x})$, depending on the form of f. Now, since $t[\mathbf{y}]$ is a symmetry of W, it is also a symmetry of L. Thus, $\mathbf{y}$ is another point of L.

See Equations 6a, 6b of the Isometry Toolkit.

We have shown that, for any $\mathbf{x} \in L$, the point $\mathbf{y} = r[\theta](\mathbf{x})$ or $\mathbf{y} = q[\theta](\mathbf{x})$ also belongs to L. Thus $r[\theta]$ or $q[\theta]$, the linear part of f, is a symmetry of L, as required. ∎

Note that we have *not* claimed that the *translation* part of f is a symmetry of L. And indeed, it need not be!

Example 2.4

In Exercise 1.1, we saw a wallpaper pattern W with a glide reflection symmetry f that was not a symmetry of any associated lattice. In the drawing of W in Figure 2.5, we have

$$f = q[(0,1),(0,0),\pi/2]$$
$$= t[(0,1)]\,q[\pi/2].$$

See Equation 14 of the Isometry Toolkit.

But neither the translation part, $t[(0,1)]$, nor the linear part, $q[\pi/2]$, is a symmetry of W. In fact, $t[(0,1]$ is not a symmetry of either W or L.

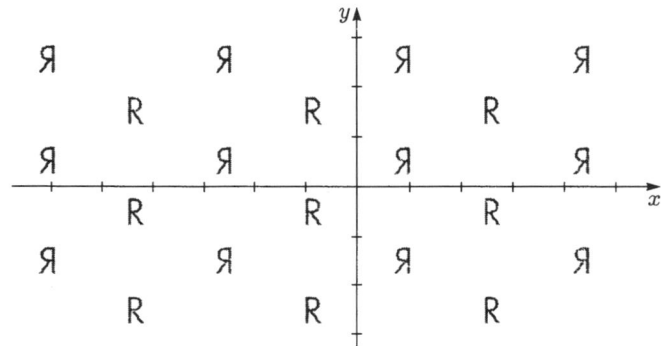

Figure 2.5

The fact that wallpaper patterns can have symmetries that are not possessed by any associated lattice may seem a matter of some concern. Does this mean that a wallpaper pattern can have some new kind of symmetry which we have not met in our study of lattices? We shall see that the answer is negative. The symmetries of a wallpaper pattern will all occur as symmetries of *some* lattice, but these symmetries will not necessarily be those of a lattice associated with the pattern. We shall look into this matter shortly.

We now introduce what is called the *point group* of a wallpaper pattern. Given a wallpaper pattern W and any point O in $\mathbb{R}^2$, we shall be looking at those isometries that fix O and that, when composed with some translation, yield a symmetry of W. The following theorem shows that they form a group.

Theorem 2.3

Let W be a wallpaper pattern with origin O. Let H be the set of all isometries h that fix O and are such that th is a symmetry of W, for *some* translation t. Then H is a group, called the **point group**, of W.

Proof

The identity element e is in H, since $t[\mathbf{0}]\,e$ is certainly a symmetry of W.

Let h and h' be elements of H. Then there are translations $t[\mathbf{x}]$ and $t[\mathbf{y}]$ such that $t[\mathbf{x}]\,h$ and $t[\mathbf{y}]\,h'$ are symmetries of W. In the composite $t[\mathbf{x}]\,h\,t[\mathbf{y}]\,h'$ we may replace $h\,t[\mathbf{y}]$ by $t[\mathbf{z}]\,h$, where $\mathbf{z} = h(\mathbf{y})$, and get the expression $(t[\mathbf{x}]\,t[\mathbf{z}])(hh')$. The composite of two symmetries of W is certainly a symmetry of W, and $t[\mathbf{x}]\,t[\mathbf{z}]$ is a translation. We also see that hh' fixes O, so hh' is an element of H. Thus H is closed.

See Equation 6 of the Isometry Toolkit.

Let h be in H. Then there is a translation $t[\mathbf{x}]$ such that $t[\mathbf{x}]\,h$ is a symmetry of W. The inverse of $t[\mathbf{x}]\,h$ will be a symmetry of W, and we can express it as

$$h^{-1}\,(t[\mathbf{x}])^{-1} = h^{-1}\,t[-\mathbf{x}] = t[\mathbf{w}]\,h^{-1},$$

where $\mathbf{w} = h^{-1}(-\mathbf{x})$. Since $t[\mathbf{w}]$ is a translation and h^{-1} fixes O, it follows that h^{-1} is in H.

We have shown that H satisfies the axioms for a subgroup (of the group Γ of all plane isometries). Thus H is a group. ∎

In the above, we have specified an origin O. The group H therefore consists of certain isometries which fix O. The choice of origin does not matter at all as far as the structure of H is concerned. It is not difficult to see that a different choice of origin will give us a group isomorphic to H.

Exercise 2.4

Find the point group of the wallpaper pattern illustrated in Figure 1.5.

Exercise 2.5

Find the point group of the wallpaper pattern illustrated in Figure 1.3.

Notice that the point group of a wallpaper pattern W may contain elements that are not symmetries of W. They are, however, symmetries of the associated lattice based at O. This follows from Theorem 2.2.

There is an obvious way of mapping the symmetry group G of a wallpaper pattern to its point group H. For each $g \in G$ there is a unique $h \in H$ such that $g = th$ for some translation t. This leads to the following result.

Theorem 2.4

Let W be a wallpaper pattern with symmetry group G, translation group T_2 and point group H. Then the mapping $G \to H$ given by

$$g \mapsto h \text{ if and only if } g = th, \text{ for some translation } t,$$

is a homomorphism of G onto H whose kernel is the group T_2. The group T_2 is therefore a normal subgroup of G.

For the remainder of this unit, we shall always use G for the symmetry group $\Gamma(W)$ of a given wallpaper pattern W, T_2 for the translation group $\Delta(W)$, and H for the point group.

Proof

Suppose $g_1 = t_1 h_1$ and $g_2 = t_2 h_2$ for some translations t_1 and t_2. Then

$$\begin{aligned} g_1 g_2 &= t_1 h_1 t_2 h_2 \\ &= t_1 \left(h_1 t_2 h_1^{-1} \right) h_1 h_2. \end{aligned}$$

Since $h_1 t_2 h_1^{-1}$ is a translation, so is $t_1 \left(h_1 t_2 h_1^{-1} \right)$ and it follows that $h_1 h_2$ belongs to H. The mapping is therefore a homomorphism. From the definition of H, the mapping is onto H. The kernel consists of those elements g in G for which $g = te = t$. These are the translations that belong to G. Hence the kernel is T_2. ∎

From our knowledge of the symmetries of various lattices, it is easy to find all the possible point groups of wallpaper patterns. They must be subgroups of the stabilizer $\mathrm{Stab}(\mathbf{0})$ for the associated lattice based at O.

In Section 4 of *Unit GE3*, we found these stabilizers (sometimes in the main text and sometimes as a solution to an exercise). They are collected in Table 2.1.

Lattice	$\mathrm{Stab}(\mathbf{0})$
Parallelogram	C_2
Rectangular	$D_2 \ (=V)$
Rhombic	$D_2 \ (=V)$
Square	D_4
Hexagonal	D_6

Table 2.1

Lattices are particular kinds of wallpaper patterns and these groups $\mathrm{Stab}(\mathbf{0})$ are their point groups. Notice that the rectangular lattice and the rhombic lattice have the same point group D_2. This shows that the point group does not determine the type of lattice.

By inspecting the subgroups of these groups, we obtain the following theorem.

Theorem 2.5

If W is a wallpaper pattern, then its point group H is one of the groups

$$C_1, C_2, C_3, C_4, C_6, D_1, D_2, D_3, D_4 \text{ or } D_6.$$

You should recall from Section 5 of *Unit GE2* that C_2 and D_1, though isomorphic, are geometrically distinct.

As you will see in later sections, all these groups actually occur as point groups for wallpaper patterns. Notice that the point group will be cyclic if and only if the wallpaper pattern has no reflections or glide reflections. The presence of any indirect symmetry ensures that the point group is dihedral.

When you study all the different types of wallpaper patterns, you will see that each of the groups C_1, C_2, C_3, C_4, C_6 and D_6 are the point groups of exactly one type of wallpaper pattern. You will find that there are three distinct patterns with point group D_1, four patterns with point group D_2, two patterns with point group D_3 and two patterns with point group D_4.

2.3 Direct and indirect symmetries

Just as with friezes and lattices, the direct symmetries of a wallpaper pattern W form a group G^+ which contains T_2 (the translation group) and is contained in G:

$$T_2 \subseteq G^+ \subseteq G.$$

The group T_2 is the translation group of an associated lattice $L = L(\mathbf{a},\mathbf{b})$:

$$T_2 = \langle t[\mathbf{a}], t[\mathbf{b}] : t[\mathbf{a}]\, t[\mathbf{b}] = t[\mathbf{b}]\, t[\mathbf{a}]\rangle.$$

As in *Unit GE3*, we usually find it convenient to abbreviate $t[\mathbf{a}]$ to t_a and $t[\mathbf{b}]$ to t_b.

As with friezes, it is possible that the translation group may be the whole of the direct symmetry group; you saw in Exercise 1.5 that the wallpaper type $p1$ is an example.

If T_2 is not the whole of G^+, then the remaining elements of G^+ are the non-trivial rotations of W. The following theorem shows that G^+ is generated by T_2 and at most one rotation.

Theorem 2.6

Let W be a wallpaper pattern, and let G, G^+ and T_2 have the usual meanings.

(a) If G has no non-trivial rotations, then $G^+ = T_2 = \langle t_a, t_b\rangle$.

(b) If G has a non-trivial rotation, and r is a rotation of maximum order, then

$$G^+ = \langle T_2, r\rangle = \langle t_a, t_b, r\rangle.$$

Proof

(a) From Theorem 5.1 of *Unit IB1*, any direct symmetry of W must be a non-trivial rotation or a translation. Therefore, if there are no non-trivial rotations, then $\Gamma^+(W) = \Delta(W)$; and since W is a wallpaper pattern, its translation group is of the form $T_2 = \langle t_a, t_b\rangle$.

(b) Let us choose the origin O as the centre of the rotation r of maximum order; suppose this order is n. Thus we may take $r = r[2\pi/n]$. Then the order of every rotation $r[\mathbf{c}, \theta]$ of W must divide n.

Now write $r[\mathbf{c}, \theta]$ in standard form, as $t\, r[\theta]$ where t is a translation and $r[\theta]$ has centre O. Then $r[\theta]$ has the same order as $r[\mathbf{c}, \theta]$ — that is, the order of $r[\theta]$ divides that of r. Thus,

$$r[\theta] = r^k \text{ for some integer } k,$$

and so $r[\theta]$ is a symmetry of W. Hence t is also a symmetry of W, and $r[\mathbf{c}, \theta] = t\, r^k$.

Thus every element of G^+ is either a translation or a translation composed with a power of r. That is, every element of G^+ is a translation composed with a (possibly zero) power of r. In other words, $G^+ = \langle T_2, r\rangle$. ∎

If the order of $r[\mathbf{c}, \theta]$ is m, then $\langle r[\mathbf{c}, \theta]\rangle$ contains $r[\mathbf{c}, 2\pi/m]$, and $r[\mathbf{c}, 2\pi/m]\, r^{-1}$ has order equal to $\mathrm{lcm}(m, n)$. If m does not divide n then this lcm is greater than n.

The group T_2 is a normal subgroup of G, by Theorem 2.4, and is therefore also a normal subgroup of G^+.

Exercise 2.6

Write out explicitly the elements of the quotient group G^+/T_2.

Let us now examine the indirect symmetries of a wallpaper pattern W more closely. The situation is essentially the same as for the indirect symmetries of a lattice, which we considered in detail in Subsections 2.1 and 5.2 of *Unit GE3*.

Reflection and glide axes

As we saw in Subsection 2.1 of *Unit GE3*, any indirect isometry is either a reflection or a glide reflection. If it is a glide reflection, then it has a translation component and a reflection component, and (unlike the translation and linear parts) the components do not depend on the choice of origin.

Now let W be any wallpaper pattern, and let f be an indirect symmetry of W. The axis of f is the axis of the reflection component of f; this axis is a reflection axis of W if the reflection component of f is a symmetry of W, otherwise it is a glide axis of W.

Exercise 2.7

Figure 2.6 is a repeat of Figure 2.4. For each of the following indirect symmetries of W (you need *not* check that they are symmetries), determine whether its axis is a reflection axis or a glide axis of W.

Hint Recall that $T_2 = \langle t[(4,0)], t[(0,4)] \rangle$.

(a) $f = t[(4,2)]\, q[0]$
(b) $g = t[(2,0)]\, q[\pi/2]$
(c) $h = t[(0,4)]\, q[\pi/4]$

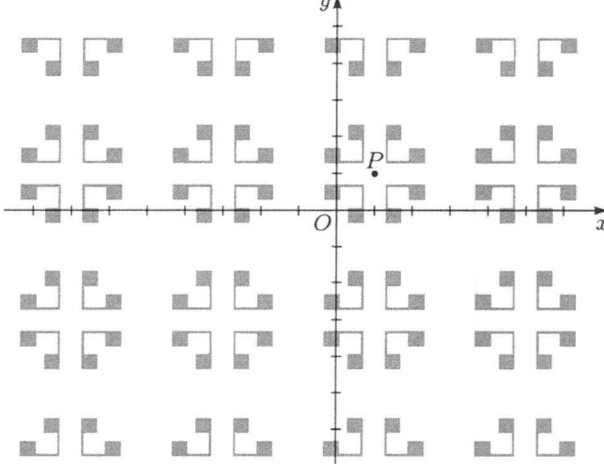

Figure 2.6

Rectangular and rhombic symmetries

You saw in Subsection 5.2 of *Unit GE3* that any indirect symmetry of a plane lattice is either rectangular or rhombic, but not both. The definition of these types of symmetry extends very easily to symmetries of a wallpaper pattern W. We simply choose an origin O for W, and let L be the associated lattice based at O. Then an indirect symmetry f of W is **rectangular** or **rhombic** depending on whether its linear part is a rectangular or a rhombic symmetry of L.

We know from Theorem 2.2 that the linear part of f is indeed a symmetry of L, and from Theorem 5.1 of *Unit GE3* that it is either rectangular or rhombic but not both.

Exercise 2.8

Classify each indirect symmetry of Exercise 2.7 as rectangular or rhombic.

One final piece of information is needed before we go on to classify all types of wallpaper patterns. You may have noticed in your study of *Unit GE3* that the rhombic lattice has, in each reflection direction, alternating reflection and glide axes. The same is true of the hexagonal lattice (all of whose indirect symmetries are rhombic). For the square lattice, however, all the axes in the horizontal and vertical directions (i.e. those of the rectangular symmetries) are axes of reflection, while the axes in the diagonal (i.e. rhombic) directions are alternately reflection and glide axes.

We now prove a theorem that extends this observation to all wallpaper patterns.

Theorem 2.7 Rectangular and rhombic axes

Let W be a wallpaper pattern, and let f be any indirect symmetry of W.

(a) If f is rectangular, then
   either all the axes of W parallel to the axis of f are reflection axes
   or all the axes of W parallel to the axis of f are glide axes.

(b) If f is rhombic, then the axes of W parallel to the axis of f are alternately reflection and glide axes.

Proof

Choose the origin O to lie on the axis of f.
Let L be the associated lattice based at O, and let q be the linear part of f.
We now consider Cases (a) and (b).

This proof is optional material.

Case (a) f is rectangular

Then there is a basis $\{\mathbf{a}, \mathbf{b}\}$ of L such that

$$q(\mathbf{a}) = \mathbf{a}, \quad q(\mathbf{b}) = -\mathbf{b}.$$

We now choose the x-axis to lie in the direction of $\mathbf{a}$ and the y-axis in the direction of $\mathbf{b}$. Thus $q = q[0]$, and as q is the linear part of f, there must be numbers x and y (not necessarily integers) such that

$$f = t[x\mathbf{a} + y\mathbf{b}]\, q[0]$$
$$= t[(xa, yb)]\, q[0] \quad \text{where } a = ||\mathbf{a}||,\ b = ||\mathbf{b}||.$$

But we chose O to lie on the axis of f, and so $y = 0$. Moreover, using the Isometry Toolkit, we obtain

$$f^2 = t[2x\mathbf{a}],$$

as you may check for yourself.

Now f^2 must be a symmetry of W, and so $2x$ is an integer.

The set S of all indirect symmetries of W with axes parallel to that of f is just the set of composites of f with translations of L:

$$S = \{t[n\mathbf{a} + m\mathbf{b}]\, t[(xa, 0)]\, q[0] : n, m \in \mathbb{Z}\}$$
$$= \{t[((n+x)a, mb)]\, q[0] : n, m \in \mathbb{Z}\}.$$

If x is an integer, we may rewrite this as

$$S = \{t[(n'a, mb)]\, q[0] : n', m \in \mathbb{Z}\}.$$

$n' = n + x$

Thus, every choice of m gives an axis parallel to the x-axis and passing through the point $(0, \tfrac{1}{2}mb)$. Choosing $n' = 0$ gives a reflection in this axis, and so all the axes are reflection axes.

If x is not an integer, we may write

$$S = \{t[(n'' + \tfrac{1}{2})a, mb)]\, q[0] : n'', m \in \mathbb{Z}\},$$

$n'' = n + x - \tfrac{1}{2}$

and this time all the axes are glide axes.

Case (b) *f is rhombic.*

Then there is a basis $\{\mathbf{a}, \mathbf{b}\}$ of L such that

$$q(\mathbf{a}) = \mathbf{b}, \quad q(\mathbf{b}) = \mathbf{a}.$$

This time, let $\mathbf{a} + \mathbf{b} = \mathbf{c}$, $\mathbf{a} - \mathbf{b} = \mathbf{d}$, and choose the x-axis and y-axis in the directions of $\mathbf{c}$ and $\mathbf{d}$ respectively. Again, $q = q[0]$, and (arguing as in Case (a)) there must therefore be a number x, where $2x$ is an integer, such that

$$f = t[x\mathbf{c}]\, q[0].$$

Thus the set S of all indirect symmetries of W with axes parallel to that of f is

$$\begin{aligned}
S &= \{t[n\mathbf{a} + m\mathbf{b}]\, t[x\mathbf{c}]\, q[0] : n, m \in \mathbb{Z}\} \\
&= \{t[\tfrac{1}{2}n(\mathbf{c} + \mathbf{d}) + \tfrac{1}{2}m(\mathbf{c} - \mathbf{d}) + x\mathbf{c}]\, q[0] : n, m \in \mathbb{Z}\} \\
&= \{t[(\tfrac{1}{2}(n + m) + x)\mathbf{c} + \tfrac{1}{2}(n - m)\mathbf{d}]\, q[0] : n, m \in \mathbb{Z}\}.
\end{aligned}$$

Different choices of $n - m$ give different axes (all parallel to $\mathbf{c}$). Now $n - m$ and $n + m$ always have the same parity (that is, they are both even or they are both odd). If x is an integer, then the even choices for $n - m$ and $n + m$ give reflection axes and the odd choices give glide axes.

If x is not an integer, then the odd choices for $n - m$ and $n + m$ give reflection axes, and the even choices give glide axes.

In either case, therefore, the reflection and glide axes alternate. ∎

Exercise 2.9

Suppose f and g are two indirect symmetries of a wallpaper pattern W, with axes at right angles. Show that if f is rectangular, then so is g, and if f is rhombic, then so is g.

Theorem 2.7 is more powerful than it looks! It allows us to deduce the main result of the unit quite easily.

Theorem 2.8 Seventeen wallpaper groups

There are just seventeen types of wallpaper pattern (or, equivalently, of wallpaper group).

Proof

Let W be a wallpaper pattern. By Theorem 2.4, its point group H is one of the groups C_i, D_i ($i = 1, 2, 3, 4, 6$).

Case 1 *H is a cyclic group*

In this case, there are no indirect symmetries, and $G = G^+$.

If $H = C_1$, then by Theorem 2.6,

$$G = G^+ = T_2.$$

If $H = C_2$, then (again by Theorem 2.6)

$$G = G^+ = \langle t_a, t_b, r \rangle$$

where $r = r[\pi]$. Thus

$$rt_a = -t_a r, \quad rt_b = -t_b r.$$

This pair of relations completely defines G.

If $H = C_3$, C_4 or C_6, let r be a rotation of maximum order. By Theorem 2.2, r is a rotation of the associated lattice based at O, which is hexagonal in the case of C_3 or C_6 or square in the case of C_4. In each case, this completely determines the relations between t_a, t_b and r (see Subsections 4.4 and 4.5 of *Unit GE3*), and, by Theorem 2.6, $G = G^+$ is completely determined.

Case 2 H is a dihedral group

If $H = D_1$, there are axes in just one direction. If f is any indirect symmetry, it may be rectangular or rhombic. If rectangular, then there are two possibilities: the axes are all reflection axes, or they are all glide axes. If f is rhombic, there is only one possibility: the axes alternate.

Thus, $H = D_1$ yields just three types of wallpaper group.

If $H = D_2$, there are axes in two directions at right angles to each other. By Exercise 2.9, both directions are rectangular or both are rhombic.

If both directions are the directions of rectangular axes, then there are three possibilities:
- the axes in both directions are reflection axes;
- the axes in both directions are glide axes;
- there are reflection axes in one direction and glide axes in the other.

If both directions are the directions of rhombic axes, then reflection and glide axes alternate in each direction.

Thus, $H = D_2$ yields just four types of wallpaper group.

If $H = D_3$, then the associated lattice L must be hexagonal. Therefore, all axes are rhombic. There are three directions of axes, inclined at angles of $\pi/3$ to each other, and they must be three of the six directions of axes of L.

There are just two possibilities: either they are the axes lying *along* the directions of **a**, **b** and **a** − **b** where $\{\mathbf{a}, \mathbf{b}\}$ is a reduced basis of L, or they are the axes *perpendicular* to these directions.

Thus, $H = D_3$ yields just two types of wallpaper group.

If $H = D_4$, then the associated lattice L must be square. Let $\{\mathbf{a}, \mathbf{b}\}$ be a reduced basis for L.

There are axes in four directions, which must be the two (rectangular) directions of **a** and **b** and the two (rhombic) directions of **a** + **b** and **a** − **b**. The two sets of rhombic axes must each be alternately reflection and glide axes, while the two sets of rectangular axes must all be reflection axes or all glide axes. (It is impossible to have all reflection axes in the **a** direction, say, and all glide axes in the **b** direction, because if q is a reflection and r is a rotation of order 4, then rqr^{-1} is a reflection with axis at right angles to that of q.)

Thus, $H = D_4$ yields just two types of wallpaper group.

Finally, if $H = D_6$, then the associated lattice L must be hexagonal. There are six directions of axes; all are rhombic, and so all have alternating reflection and glide axes. Thus, $H = D_6$ yields just one type of wallpaper group.

We have been through all possibilities, of which there are just seventeen. ∎

3 PARALLELOGRAM AND RECTANGULAR PATTERNS

3.1 Parallelogram patterns: $p2$ and $p1$

In this subsection we consider patterns for which the associated lattices can be of parallelogram type. This means that a design for such a pattern may be drawn on a basic parallelogram of any parallelogram lattice. It follows that these patterns have no indirect symmetries.

Pattern type $p2$

According to the notation introduced in Subsection 1.2, a pattern of type $p2$ has rotations of order 2 but none of higher order. There are also no reflections or glide reflections. A typical example is the parallelogram lattice itself.

We have already seen examples of patterns of this type. The designs shown in Figure 2.1 are all of type $p2$ and so is pattern (a) of Figure 2.2. Notice that, from among these designs, only one of them has an associated lattice which is of parallelogram type.

Figure 3.1 shows another example of a pattern of this type.

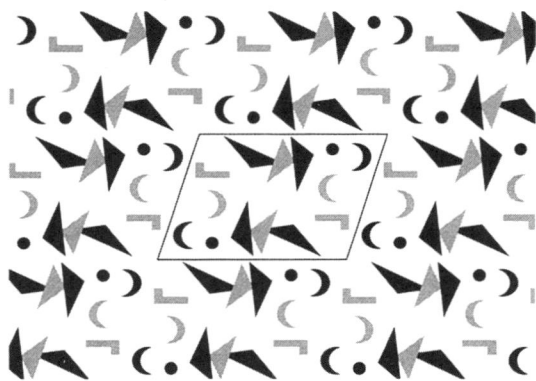

Figure 3.1

In Figure 3.2, we show a pattern whose basic design consists of two flags and we have shaded a generating region consisting of a triangle.

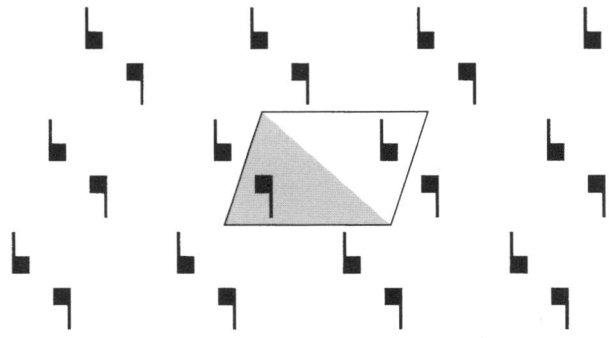

Figure 3.2

Since the group G of all symmetries of this pattern is isomorphic to the group of symmetries of a parallelogram lattice, it follows that G can be written as

$$G = \{x, xr : x \in T_2;\ r^2 = e,\ rt_a = t_a^{-1}r,\ rt_b = t_b^{-1}r\}$$

or, more compactly, as

$$G = \{xy : x \in T_2,\ y \in C_2;\ rt_a = t_a^{-1}r,\ rt_b = t_b^{-1}r\},$$

See Subsection 4.1 of *Unit GE3*.

where T_2 is the translation group and $C_2 = \{e, r\}$ is the cyclic group of order 2.

Exercise 3.1

Find the point group of the wallpaper group $p2$.

Pattern type $p1$

A pattern of type $p1$ is the simplest of all. It can be drawn so that its associated lattice is of parallelogram type but it does not possess any of the rotations of order 2. Here, in Figure 3.3, is an example.

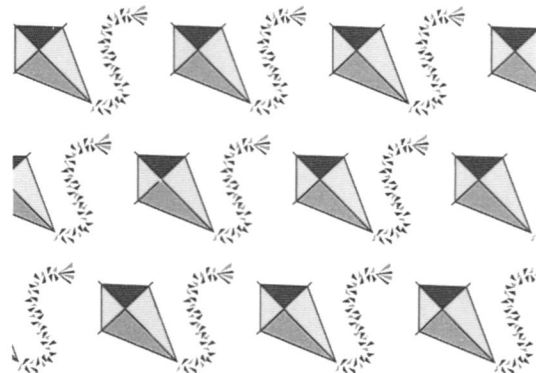

Figure 3.3

Since there are no rotations, reflections or glide reflections, a generating region for a pattern of type $p1$ will be any basic parallelogram. No smaller region will suffice. In Figure 3.4, we show a simpler example which has a basic parallelogram containing one flag. This parallelogram has been shaded to indicate that it is a generating region.

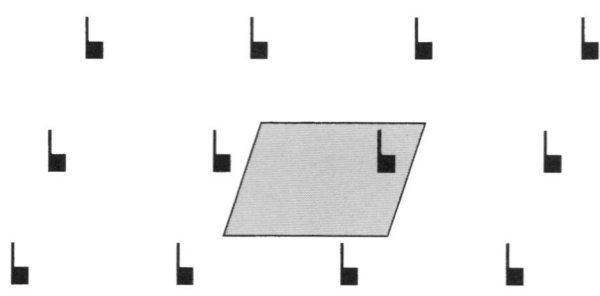

Figure 3.4

The group of symmetries G of a pattern of type $p1$ is simply the group T_2 of translations.

Exercise 3.2

Find the point group of the wallpaper group $p1$.

The wallpaper patterns in the next five subsections are based on a rectangular lattice. All the patterns that we shall be studying here have indirect symmetries of rectangular type. There will be none of rhombic type.

3.2 Pattern type $p2mm$

A pattern of type $p2mm$ is typified by the rectangular lattice itself. A more striking example is shown in Figure 3.5.

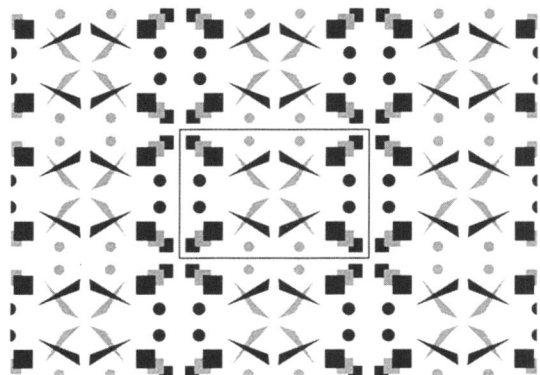

Figure 3.5

The two ms show that there are reflection axes in both the horizontal and vertical directions. Let q be a reflection with a horizontal axis and let q' be a reflection with a vertical axis. If we choose the origin to be the intersection of these axes, then $r = qq'$ will be a rotation of order 2 with centre $\mathbf{0}$. The stabilizer of the point $\mathbf{0}$ will be the Klein group $V = \{e, r, q, rq\}$, also known as the dihedral group D_2. Writing the translation group as $T_2 = \langle t_a, t_b \rangle$ with t_a horizontal and t_b vertical, the full group of symmetries (as in Subsection 4.2 of *Unit GE3*) is

$$G = \{xy : x \in T_2, y \in D_2;\; rt_a = t_a^{-1}r,\; rt_b = t_b^{-1}r,\; qt_a = t_aq,\; qt_b = t_b^{-1}q\}.$$

In Figure 3.6, we show a simpler example where a basic rectangle contains four flags.

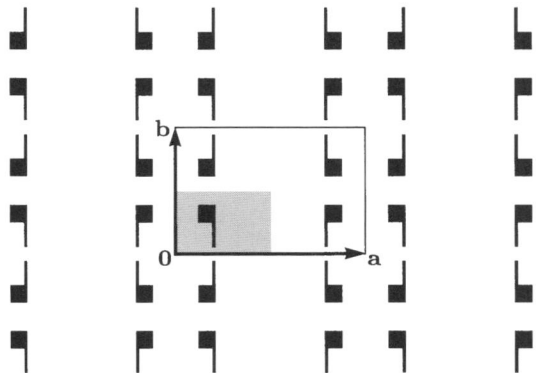

Figure 3.6

The shaded rectangle is a generating region for the pattern, as you may readily check. A pattern of type $p2mm$ will have four orbits of 2-centres. With the origin O located at the bottom left-hand corner of the rectangle, these four orbits are represented by the points $\mathbf{0}$, $\tfrac{1}{2}\mathbf{a}$, $\tfrac{1}{2}\mathbf{b}$ and $\tfrac{1}{2}\mathbf{a} + \tfrac{1}{2}\mathbf{b}$.

The stabilizer of each 2-centre will consist of the identity, a rotation of order 2 and reflections in perpendicular axes which pass through the 2-centre. Each of these stabilizers is the dihedral group D_2 of order 4.

There are four orbits of reflection axes. The lines through **0** which contain **a** and **b**, respectively, will lie in distinct orbits. The lines parallel to these and passing through the point $\frac{1}{2}\mathbf{a} + \frac{1}{2}\mathbf{b}$ will be representatives of two more orbits. As we saw in *Unit GE3*, the stabilizers of these axes are frieze groups of Type 6.

Exercise 3.3

In terms of r, q, t_a and t_b, find the symmetries which map the given generating region into the basic rectangle shown.

Exercise 3.4

Find the point group of the wallpaper group *p2mm*.

> Friezes of Type 6 are denoted by f_{vh} in Section 4 of *Unit IB3*, or by *pmm*2 in the international notation for friezes. Here, the order of rotation (2) is the *last* symbol; this is simply a device to distinguish the frieze from the wallpaper notation.

3.3 Pattern type *p2mg*

In Figure 3.7, we show part of a wallpaper pattern whose type is different from any patterns that we have seen so far. Some horizontal lines are reflection axes but there are no reflections with vertical axes. Some of the vertical lines, however, are glide axes.

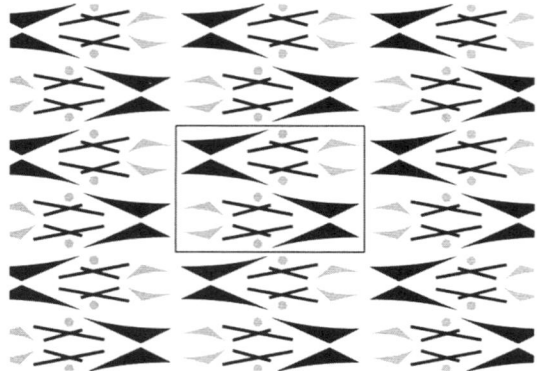

Figure 3.7

It is easier to study the simpler example shown in Figure 3.8. We have chosen a basic rectangle and shaded a generating region.

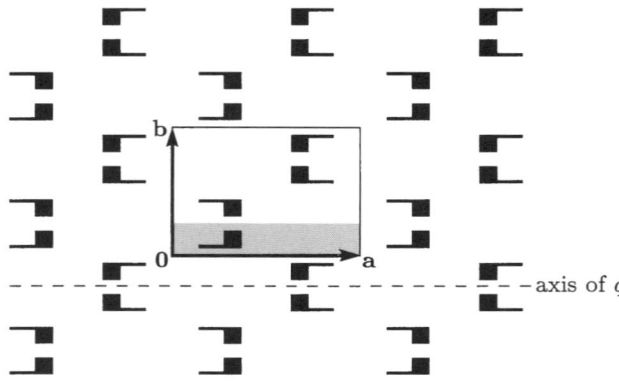

Figure 3.8

For this basic rectangle, the vertices are 2-centres. Let us introduce some notation. We choose the origin $\mathbf{0}$ to be the 2-centre at the bottom left-hand corner of the rectangle and let t_a and t_b be the usual horizontal and vertical translations. Let r be the rotation of order 2 with centre $\mathbf{0}$ and let q be the reflection in a horizontal axis which passes through the point $-\frac{1}{4}\mathbf{b}$. Then $g' = rq$ will be a glide reflection whose axis passes through $\mathbf{0}$ and is directed upwards.

We could equally well have chosen the axis of q to pass through the point $\frac{1}{4}\mathbf{b}$, and hence through the basic rectangle. The reason for our choice is to simplify the relations in our description of G.

The notation for a pattern of this type is $p2mg$. The letter p and the number 2 have the same significance as in the type $p2mm$, as has the first letter m — namely that, choosing the reference direction to point downwards as in Subsection 1.2, there are reflection axes perpendicular to the reference direction.

The new feature of this notation is that the final letter is g. This refers to the fact that the axes in the other direction are glide axes.

Exercise 3.5

Find representatives of the orbits of 2-centres and specify their stabilizers.

All the symmetries of a wallpaper pattern of type $p2mg$ can be expressed in terms of r, q and translations in T_2. The subgroup of direct symmetries is

$$G^+ = \{x, xr : x \in T_2;\ r^2 = e,\ rt_a = t_a^{-1}r,\ rt_b = t_b^{-1}r\}.$$

We also have the relations

$$q^2 = e, \quad qt_a = t_aq \quad \text{and} \quad qt_b = t_b^{-1}q.$$

So far, the relations are identical to those for the pattern $p2mm$. It is when we look at the composite rq that we find a difference. For this pattern, the axis of q does not pass through the centre of the rotation r. This means that rq is a glide reflection rather than a reflection. In our notation above, we have

$$rq = g',$$

where $g'^2 = t_b$, so

$$(rq)^2 = t_b.$$

Contrast this with the corresponding relation $(rq)^2 = e$ for the pattern $p2mm$.

Thus, the group G is

$$G = \{x, xr, xq, xrq : x \in T_2;\ r^2 = e,\ q^2 = e,\ (rq)^2 = t_b,\ rt_a = t_a^{-1}r,$$
$$rt_b = t_b^{-1}r,\ qt_a = t_aq,\ qt_b = t_b^{-1}q\}.$$

This could be written as

$$G = \{xyz : x \in T_2,\ y \in C_2,\ z \in D_1;\ (rq)^2 = t_b,\ \text{etc.}\},$$

but the more explicit formulation is probably less confusing. The point group H for the pattern $p2mg$ is easily found. The symmetries that fix $\mathbf{0}$ are e and r so they must belong to H. Let q' be the reflection whose axis is vertical and passes through $\mathbf{0}$. Although q' is not a symmetry of the pattern, we can write it as $q' = t[-\frac{1}{2}\mathbf{b}]\,g'$. It follows that q' belongs to H. The group H consists of e, r, q' and rq'. It is the group $D_2\ (= V)$ of order 4.

Exercise 3.6

Find the translation component of the glide reflection g'.

You will notice that the glide reflection g' is a symmetry of the pattern but not of any associated lattice, and you may recall our earlier claim that the symmetry group of a wallpaper pattern is a subgroup of some lattice group. In this case, it is not a subgroup of the group of symmetries of any associated lattice, so we need to consider another lattice. Let us take the lattice

$$L' = L(\mathbf{a}, \tfrac{1}{2}\mathbf{b})$$

whose translations form the group

$$T_2' = \langle t[\mathbf{a}], t[\tfrac{1}{2}\mathbf{b}] \rangle.$$

The symmetry group of this lattice will contain the translation $t[\tfrac{1}{2}\mathbf{b}]$ and the reflection q', so it will contain the glide reflection $g' = t[\tfrac{1}{2}\mathbf{b}]\, q'$. It will also contain T_2 and the rotation r and hence it must contain all the symmetries of the pattern.

In Figure 3.9 we show a basic rectangle for the lattice L'. The glide reflection g' is a non-essential glide reflection of L', whereas it is an essential glide reflection of W.

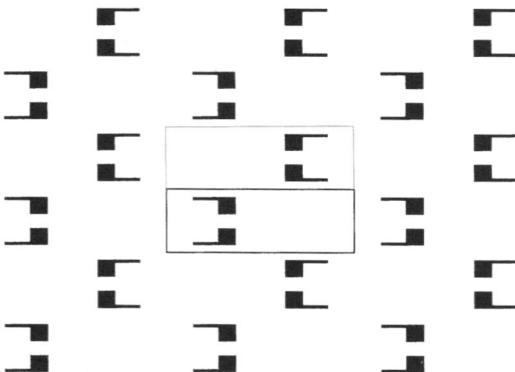

The basic rectangle for L' is the smaller one, drawn with solid lines; the dotted lines form part of the boundary of the basic rectangle for W that was drawn in Figure 3.8.

Figure 3.9

3.4 Pattern type $p2gg$

Another type of pattern of rectangular type is denoted by $p2gg$. From the second symbol of the notation, it has rotations of order 2. The third symbol g tells us that it has glide axes perpendicular to the reference direction (which points vertically down, as before). Finally, the fourth symbol g tells us that there are glide axes parallel to the reference direction. There are no axes of reflection. Figure 3.10 depicts such a wallpaper pattern.

Figure 3.10

Once again, it is much easier to study a pattern of the same type but with a simpler basic design. In Figure 3.11, we display a basic rectangle and a generating region from which we may construct the whole pattern.

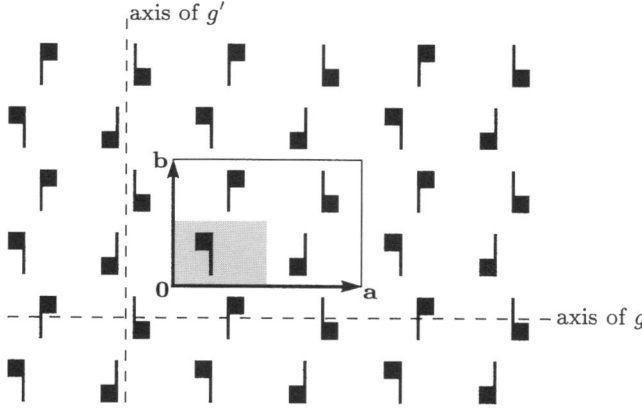

Figure 3.11

The only indirect symmetries are glide reflections in horizontal and vertical axes. Let us take our origin O to be the 2-centre at the bottom left-hand corner of the basic rectangle and let r be the rotation of order 2 with centre O. Let g be the glide reflection whose axis is horizontal and passes through $-\frac{1}{4}\mathbf{b}$ and whose translation component is $t[\frac{1}{2}\mathbf{a}]$. Then $g' = rg$ will be a glide reflection with axis through $-\frac{1}{4}\mathbf{a}$ and with translation component $t[\frac{1}{2}\mathbf{b}]$.

Once again, the reason for choosing the axis of g so that it avoids the basic rectangle is to simplify the relations in our description of G.

The symmetries of a wallpaper pattern of type $p2gg$ can all be expressed in terms of r, g and the translations in $T_2 = \langle t_a, t_b \rangle$. As in the case of $p2mm$ and $p2mg$, we have

$$G^+ = \{x, xr : x \in T_2;\ r^2 = e,\ rx = x^{-1}r\}.$$

Now the axis of g is parallel to t_a and perpendicular to t_b so we have $gt_a = t_a g$ and $gt_b = t_b^{-1} g$. We also have the relations $g^2 = t_a$ and $(rg)^2 = t_b$ and these are enough to specify the group:

$$G = \{x, xr, xg, xrg : x \in T_2;\ r^2 = e,\ g^2 = t_a,\ (rg)^2 = t_b,\ rt_a = t_a^{-1}r,$$
$$rt_b = t_b^{-1}r,\ gt_a = t_a g,\ gt_b = t_b^{-1}g\}.$$

Exercise 3.7

Find the point group of the wallpaper group $p2gg$.

An associated lattice for this pattern will not have, among its symmetries, the glide reflections g and g', so again we need to consider another lattice. Let us take the lattice

$$L' = L(\tfrac{1}{2}\mathbf{a}, \tfrac{1}{2}\mathbf{b}).$$

We can readily check that the symmetry group of L' contains all the symmetries of our pattern.

In Figure 3.12, we show a basic rectangle for the lattice L'.

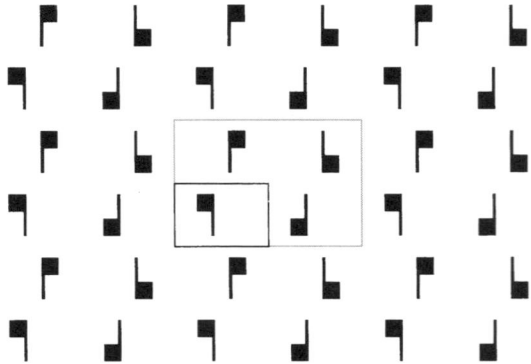

Figure 3.12

The basic rectangle for L' is the smaller one, drawn with solid lines.

3.5 Pattern type *pm*

The remaining types of pattern that are based on a rectangle have indirect symmetries but no rotations. This implies that the axes of indirect symmetries will all be parallel. The first of these has reflections and is of type *pm*. An example is given in Figure 3.13.

Strictly speaking the notation should be *p1m*, but the symbol 1 (to indicate the absence of non-trivial rotations) is normally omitted.

Figure 3.13

A simpler version is shown in Figure 3.14 where we give a basic rectangle and a generating region.

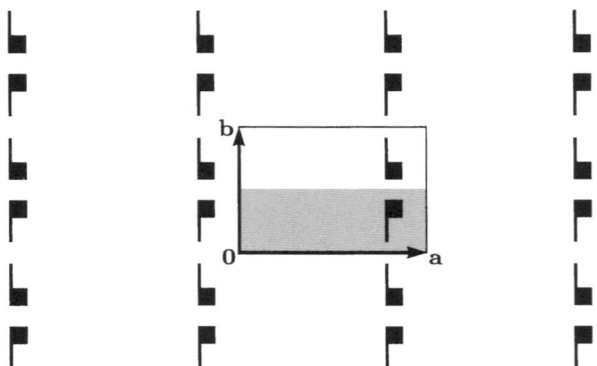

Figure 3.14

Let us choose the origin O to be the bottom left-hand corner of the rectangle shown. There is a reflection q whose axis is horizontal and passes through O. The group of symmetries of the pattern is

$$G = \{x, xq : x \in T_2;\ q^2 = e,\ qt_a = t_a q,\ qt_b = t_b^{-1} q\}$$

or, more compactly,

$$G = \{xy : x \in T_2,\ y \in D_1;\ qt_a = t_a q,\ qt_b = t_b^{-1} q\}.$$

That is to say, we take the generators of $p2mm$ except r, and the relations not involving r.

The reference direction points downwards again.

Exercise 3.8

Find the point group of the wallpaper group pm.

3.6 Pattern type pg

The last type of pattern that is based on a rectangular lattice has glide reflections in one direction but no rotations or reflections. Such a pattern is of type pg. Figure 3.15 shows one example.

Figure 3.15

A simpler example is shown in Figure 3.16, where we give a basic parallelogram and a generating region.

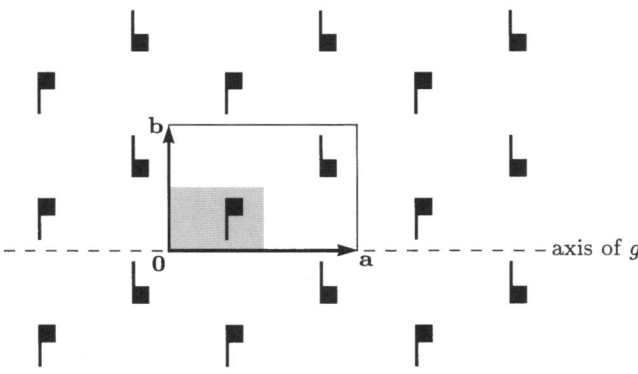

Figure 3.16

With origin O at the bottom left-hand corner of the rectangle shown, there is a glide reflection g whose axis is horizontal and passes through O. Taking the reference direction to point downwards and t_a to point horizontally, as usual, we may assume that $g^2 = t_a$. The group of symmetries of this pattern is

$$G = \{x, xg : x \in T_2;\ g^2 = t_a,\ gt_a = t_a g,\ gt_b = t_b^{-1} g\}.$$

Exercise 3.9

Find the point group of the wallpaper group *pg*.

This is another case where the symmetry group of an associated lattice does not possess all the symmetries of the pattern. We consider instead the lattice $L' = L(\frac{1}{2}\mathbf{a}, \mathbf{b})$. If q is the reflection in a horizontal axis through O, we find that $g = t[\frac{1}{2}\mathbf{a}]\,q$ and hence g is a symmetry of L'. Figure 3.17 shows a basic rectangle for the lattice L'.

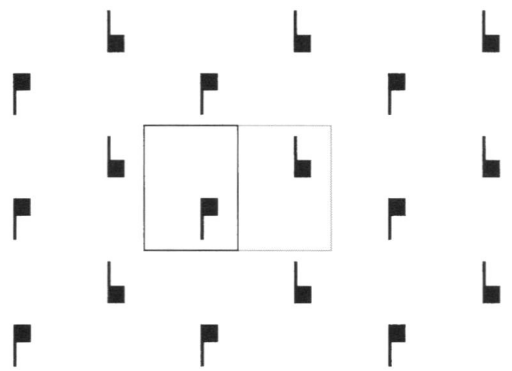

The basic rectangle for L' is the smaller one, drawn with solid lines.

Figure 3.17

In this section we have investigated two types of wallpaper patterns without indirect symmetries and five types of wallpaper patterns whose indirect symmetries are exclusively rectangular. We have gone into considerable detail; the next two sections cover the other ten types, mostly in less detail.

Mostly harmless!

4 RHOMBIC AND SQUARE PATTERNS

The first two subsections of Section 4 concern wallpaper patterns that are based on a rhombic lattice. Their indirect symmetries are all of rhombic type. We then investigate patterns whose associated lattices are square. For these, we may expect to find indirect symmetries of rectangular and rhombic types, since a square lattice has the symmetries of both a rectangular lattice and a rhombic lattice.

4.1 Pattern type $c2mm$

A pattern of type $c2mm$ has rotations of order 2 and reflections with axes in two perpendicular directions. All these reflections will be of rhombic type. The symbol c indicates that we are taking a centred cell — that is, a rectangle of twice the area of a basic parallelogram, with lattice points at the four corners and the centre. The simplest of these patterns is a rhombic lattice itself. In Figure 4.1, we show a more intricate example.

Figure 4.1

As usual, we prefer to study a pattern whose design is less detailed. In Figure 4.2, we show such a pattern together with a centred cell, a basic rhombus and a generating region.

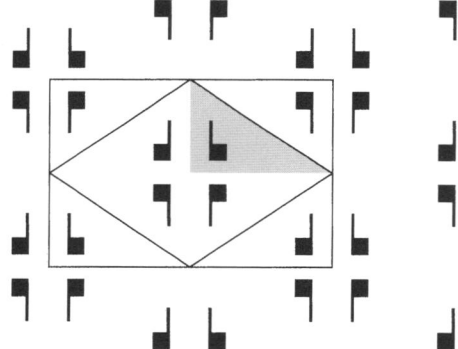

Figure 4.2

From Subsection 4.3 of *Unit GE3*, the symmetry group of W may be written as

$$G = \{xy : x \in T_2, y \in D_2; \ rx = x^{-1}r, \ qt_a = t_b q, \ qt_b = t_a q\}.$$

More explicitly,

$$G = \{xy : x \in T_2, y \in D_2; \ rt_a = t_a^{-1}r, \ rt_b = t_b^{-1}r, \ qt_a = t_b q, \ qt_b = t_a q\}.$$

Exercise 4.1

Find the point group of the wallpaper group $c2mm$.

4.2 Pattern type cm

For patterns of type cm, we again have a centred cell, so the reference direction, which points vertically down, is parallel to one of the diagonals of a basic rhombus. There will be reflections whose axes are parallel to the other diagonal but none parallel to the reference direction. In Figure 4.3, we show an example.

Figure 4.3

A less complicated example is shown in Figure 4.4 together with a centred cell, a basic rhombus and a generating region.

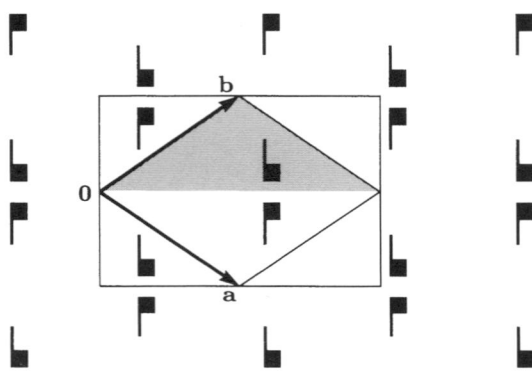

Figure 4.4

We can easily write down its group of symmetries. It will be generated by the translations and a reflection q of rhombic type, and we may assume that $qt_a = t_b q$ and $qt_b = t_a q$. The group of symmetries is

$$G = \{x, xq : x \in T_2; \ q^2 = e, \ qt_a = t_b q, \ qt_b = t_a q\}$$

or, more compactly,

$$G = \{xy : x \in T_2, \ y \in D_1; \ qt_a = t_b q, \ qt_b = t_a q\}.$$

That is to say, we take the generators of $c2mm$ except r, and the relations not involving r.

Exercise 4.2

Find the point group of the wallpaper group cm.

This completes our study of patterns of rhombic type. There are reflections of rhombic type either in two perpendicular directions or in only one direction. Notice that, by Theorem 2.7, it is not possible to have merely glide reflections.

For the rest of this section, we shall be examining patterns that have rotations of order 4.

4.3 Pattern type $p4mm$

In this subsection, we study patterns of type $p4mm$. Since the second symbol is 4, there is a rotation of order 4. This guarantees that every associated lattice is square. Now recall from Subsection 1.2 that the third symbol m asserts the presence of reflections with axes perpendicular to the reference direction. These will be reflections of rectangular type and we may let q be one of them. Because of the rotations of order 4, there are also reflection axes parallel to the reference direction. Then the fourth symbol m says that there are also reflections in diagonal axes. If q' is one of these, it will be of rhombic type. The angle between the axes of q and q' will be $\pi/4$, so the composite $q'q$ will be a rotation r of order 4 whose centre is the point of intersection of the axes of q and q'. At right angles to the axes of q and q' will be the axes of the reflections rqr^{-1} and $rq'r^{-1}$. An example of this type of pattern is shown in Figure 4.5.

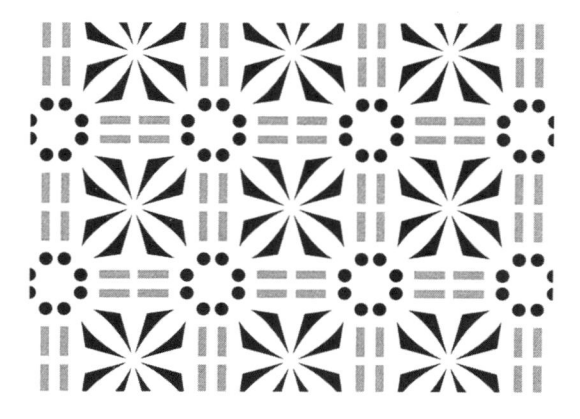

Figure 4.5

If you examine this pattern, you will see that there are many 4-centres. Through each of these 4-centres, there are reflections of rectangular type with axes in the horizontal and vertical directions. There are reflections of rhombic type with axes in both diagonal directions. The pattern also contains glide axes in both diagonal directions. The pattern has all the symmetries of the square lattice, which is itself a pattern of this type. By Subsection 4.4 of *Unit GE3*, the symmetry group can therefore be written as

$$G = \{xy : x \in T_2, y \in D_4;\ rt_a = t_b r,\ rt_b = t_a^{-1} r,\ qt_a = t_a q,\ qt_b = t_b^{-1} q\}.$$

Once again, we find it convenient to take an example where the design consists of flags. There are eight of these in a basic square. We have also shown a generating region which is a triangle whose area is one eighth that of the square. This region contains a single flag.

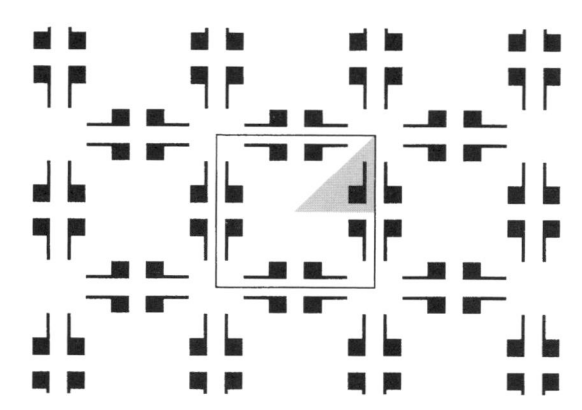

Figure 4.6

Exercise 4.3

Find the point group of the wallpaper group $p4mm$.

4.4 Pattern type $p4gm$

Another type of pattern that has rotations of order 4 is $p4gm$. Every associated lattice must be square. The third symbol g indicates that there are glide axes but no reflection axes perpendicular to the reference direction. The presence of rotations of order 4 shows that this is also true in directions parallel to the reference direction. The fourth symbol m indicates the presence of the reflection axes in the diagonal directions. (As these are the axes of rhombic symmetries, Theorem 2.7 tells us that there are glide axes in these directions as well.)

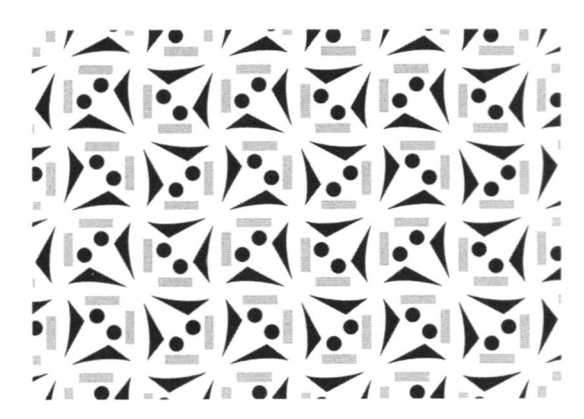

Figure 4.7

A simpler example is shown in Figure 4.8. We have chosen a basic square whose vertices are 4-centres. The shaded triangle will be a generating region for the pattern.

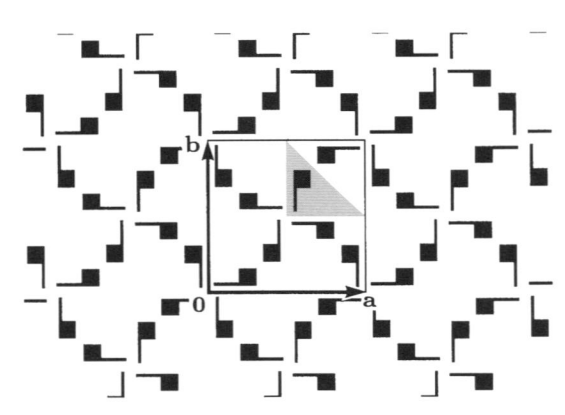

Figure 4.8

Let us take our origin O to be the bottom left-hand corner of this basic square, so that the vertices are the points $\mathbf{0}$, $\mathbf{a}$, $\mathbf{b}$ and $\mathbf{a} + \mathbf{b}$. Let r be the rotation $r[\pi/2]$. The associated lattice based at O will have a reflection q with axis parallel to $t[\mathbf{a}]$ and passing through O. Then $q' = qr$ will also be a reflection of this lattice. By inspection, we find that $q(\mathbf{a}) = \mathbf{a}$, $q(\mathbf{b}) = -\mathbf{b}$, $q'(\mathbf{a}) = -\mathbf{b}$ and $q'(\mathbf{b}) = -\mathbf{a}$.

The reflections q and q' are *not* symmetries of the wallpaper pattern itself.

Our wallpaper pattern has the glide reflection $g = t[\frac{1}{2}\mathbf{a} + \frac{1}{2}\mathbf{b}]\, q$ with axis parallel to the axis of q and the glide reflection $g' = t[-\frac{1}{2}\mathbf{a} + \frac{1}{2}\mathbf{b}]\, q'$ with axis parallel to the axis of q'.

Exercise 4.4

(a) Draw the square with vertices $\mathbf{0}$, $\mathbf{a}$, $\mathbf{b}$ and $\mathbf{a} + \mathbf{b}$ and show all the reflection and glide axes that meet this square.

  Mark the axes of g, g' and rg.

(b) Find the translation components of g and g'. Verify that $r = gg'$.

The group of symmetries of the pattern may be generated from $T_2 = \langle t_a, t_b \rangle$ and the symmetries r and g. To specify the group we need to find a complete set of relations involving these symmetries. It is easy to see that $r^4 = e$, $rt_a = t_b r$ and $rt_b = t_a^{-1} r$. The rest are to be found in the following exercise.

Exercise 4.5

Verify the relations $g^2 = t_a$, $gt_a = t_a g$, $gt_b = t_b^{-1} g$ and $(rg)^2 = e$.

The subgroup that fixes the origin O is $C_4 = \{e, r, r^2, r^3\}$, and we may write our group elements in one of the forms xy and xyg, where $x \in T_2$ and $y \in C_4$. The relations above will allow us to express composites of symmetries in one of these forms.

The complete symmetry group can now be written as

$$G = \{xy, xyg : x \in T_2, y \in C_4; \; rt_a = t_b r, \; rt_b = t_a^{-1} r, \; g^2 = t_a,$$
$$gt_a = t_a g, \; gt_b = t_b^{-1} g, \; (rg)^2 = e\}.$$

If you look at the generating region shown in Figure 4.8, you will see that it contains only one 4-centre. Hence the 4-centres of the pattern form a single orbit. These points are the intersections of axes of glide reflections of rhombic type. There is only one orbit of 2-centres which are not 4-centres. These occur at the intersections of axes of reflections of rhombic type. The stabilizer of a 4-centre is a cyclic group C_4 of order 4, and the stabilizer of a 2-centre that is not a 4-centre is a dihedral group D_2 of order 4.

Exercise 4.6

Find the point group of the wallpaper group *p4gm*.

A wallpaper pattern of type *p4gm* has glide reflections which are not symmetries of any associated lattice. Once again we need to consider the lattice $L' = L(\frac{1}{2}\mathbf{a}, \frac{1}{2}\mathbf{b})$. The symmetries of L' will contain q and q' and the translations $t[-\frac{1}{2}\mathbf{a} + \frac{1}{2}\mathbf{b}]$ and $t[\frac{1}{2}\mathbf{a} + \frac{1}{2}\mathbf{b}]$. It follows that the symmetry group of L' will contain g, g', r and the translations in T_2. In Figure 4.9, we show a basic square for the lattice L'.

The basic square for L' is the smaller one, drawn with solid lines.

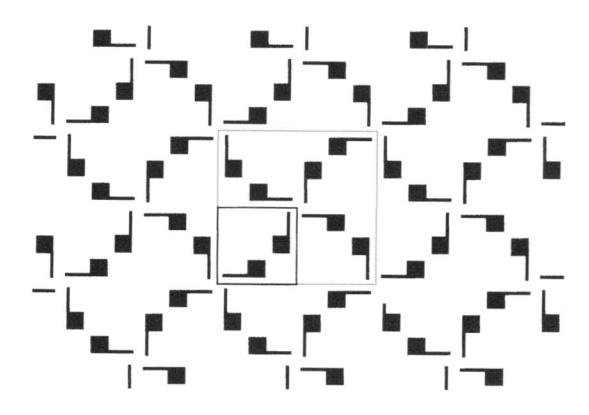

Figure 4.9

4.5 Pattern type $p4$

A pattern of type $p4$ has rotations of order 4 but no indirect symmetries. In Figure 4.10, we show an example.

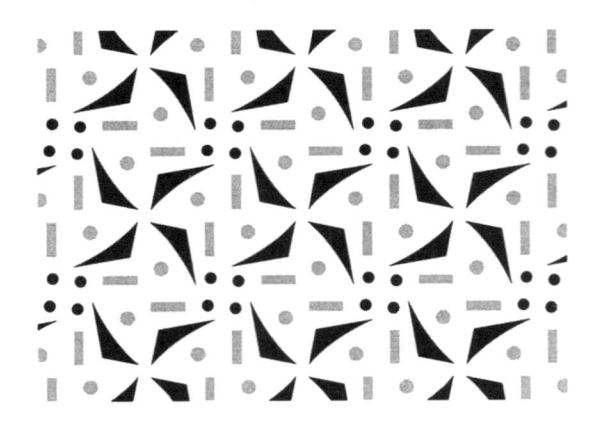

Figure 4.10

In Figure 4.11, we show a standard example with a basic square and a generating region.

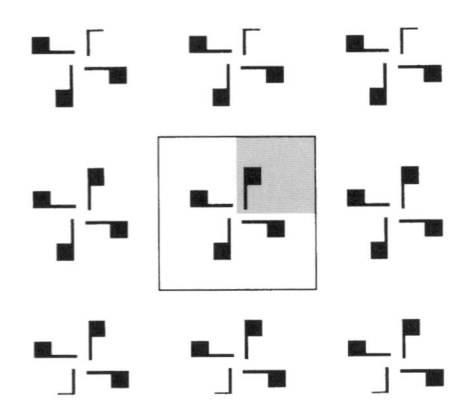

Figure 4.11

The group of symmetries of this pattern is

$$G = \{xy : x \in T_2, y \in C_4;\ rt_a = t_b r,\ rt_b = t_a^{-1} r\}.$$

This is the same as the group G^+ of direct symmetries of patterns of types $p4mm$ and $p4gm$.

Exercise 4.7

Find the point group of the wallpaper group $p4$.

We have described all the types of wallpaper patterns whose associated lattices can be square. The two types of parallelogram patterns and the five types of rectangular patterns can certainly be represented using a square as a basic parallelogram. Now, in addition, we have the two types of rhombic pattern and the three types of square pattern. You will notice that the associated lattice for a pattern of rhombic type must be either rhombic, square or hexagonal. For the square patterns, the associated lattice must be square.

5 HEXAGONAL PATTERNS

The remaining five types of wallpaper patterns are based on a hexagonal lattice. The first two types give patterns that have rotations of orders 6, 3 and 2. Patterns in the last three types possess rotations of order 3 but none of order 2.

5.1 Pattern type $p6mm$

A wallpaper pattern of type $p6mm$ has rotations of order 6. It follows that every associated lattice must be a hexagonal lattice. A basic parallelogram is therefore a rhombus whose adjacent sides subtend angles of $\pi/3$ and $2\pi/3$.

In Figure 1.6 of Subsection 1.2, the reference direction slopes downwards and to the left. We saw in the solution to Exercise 1.4 that this implies that the first m of the symbol refers to reflection axes making angles of $\pm\pi/6$ and $\pi/2$ with the x-axis (in other words, bisecting the angles between the directions of $\mathbf{a}$, $\mathbf{b}$ and $\mathbf{b} - \mathbf{a}$), while the second m refers to reflection axes along the x-axis and at angles of $\pm\pi/3$ to it (in other words, *in the* directions of $\mathbf{a}$, $\mathbf{b}$ and $\mathbf{b} - \mathbf{a}$).

The hexagonal lattice itself is an example of a wallpaper pattern of type $p6mm$. We show another example in Figure 5.1.

Figure 5.1

In Figure 5.2, we show a standard example with a basic parallelogram and a generating region.

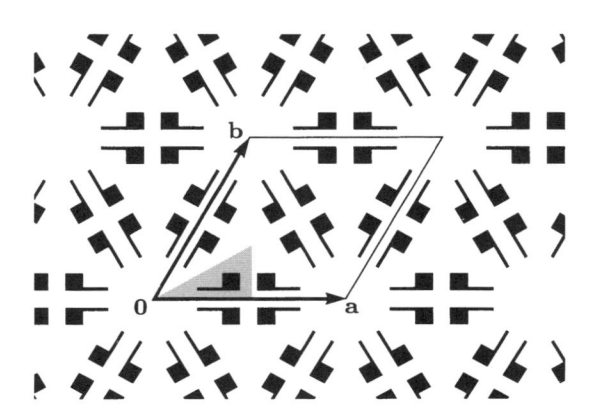

Figure 5.2

You studied this symmetry group in Subsections 3.5 and 4.5 of *Unit GE3*, under the guise of the symmetry group of the hexagonal lattice. It can be expressed as

$$G = \{xy : x \in T_2,\ y \in D_6;\ rt_a = t_b r,\ rt_b = t_a^{-1} t_b r,\ qt_a = t_a q,\ qt_b = t_a t_b^{-1} q\}.$$

Exercise 5.1

Find the point group of the wallpaper group *p6mm*.

We can see that *p6mm* is the only type of pattern that has rotations of order 6 and an indirect symmetry. This is because the hexagonal lattice does not have an orthogonal basis, and it follows that every indirect symmetry must be rhombic. Using Theorem 2.7, such a pattern must possess a reflection. All the symmetries of the hexagonal lattice can then be generated.

5.2 Pattern type *p6*

A wallpaper pattern that has rotations of order 6 but no indirect symmetries will be of type *p6*. An example is shown in Figure 5.3.

Figure 5.3

Another example is shown in Figure 5.4, where we have shown a basic rectangle and a generating region.

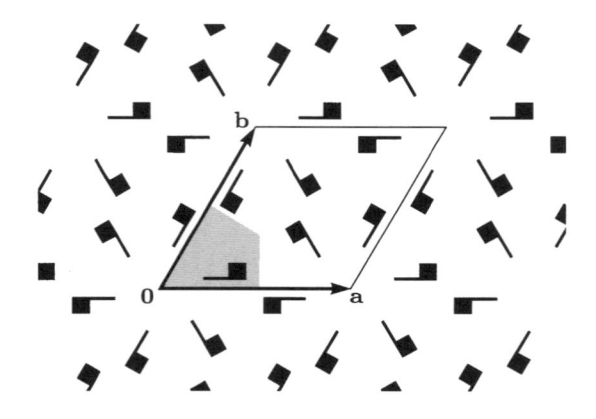

Figure 5.4

Its group of symmetries will consist entirely of direct symmetries, and we may write it as

$$G = \left\{xy : x \in T_2,\ y \in C_6;\ rt_a = t_b r,\ rt_b = t_a^{-1} t_b r \right\}.$$

Exercise 5.2

Find the point group of the wallpaper group *p6*.

5.3 Pattern type $p3m1$

A pattern of this type will have a rotation of order 3 but no rotations of order 6. It should be clear that every associated lattice must be hexagonal. The symbol m in third place indicates the presence of reflection axes perpendicular to the reference direction. If q is such a reflection and r is a rotation of order 3, the symmetries rqr^{-1} and r^2qr^{-2} will also be reflections. The reference direction is parallel to one of the sides of a basic parallelogram, so the axes of these various reflections will be perpendicular to the sides or to the short diagonal of a basic parallelogram.

In the notation $p3m1$, the fourth symbol is 1 and this indicates that there are no reflections or glide reflections in lines making an angle of $\pi/3$ with the reference direction. This implies that lines parallel to sides or the short diagonal of the basic parallelogram will not be axes of symmetries. In Figure 5.5, we show an example of a pattern of type $p3m1$.

Figure 5.5

In Figure 5.6, we show another pattern of type $p3m1$ with a basic parallelogram and a generating region.

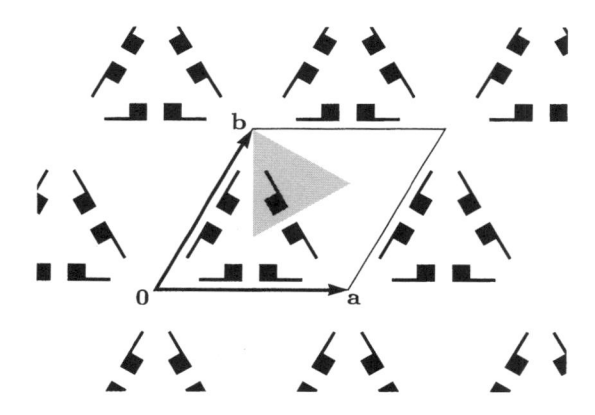

Figure 5.6

For a pattern of this type, it is natural to choose our origin O to be a 3-centre. The vertices of our basic parallelogram will then be the points $\mathbf{0}$, $\mathbf{a}$, $\mathbf{a}+\mathbf{b}$ and $\mathbf{b}$, and we take r to be the rotation $r[2\pi/3]$ with centre O.

We now need a reflection. We cannot make the same choice of axis that we made for a pattern of type $p6mm$, since the line through $\mathbf{0}$ and $\mathbf{a}$ is not an axis of reflection. Instead, we shall take, for our reflection q, the reflection whose axis is the line through $\mathbf{0}$ and $\mathbf{a}+\mathbf{b}$. We may now calculate the images of $\mathbf{a}$ and $\mathbf{b}$ under these symmetries and get

$$q(\mathbf{a}) = \mathbf{b}, \quad q(\mathbf{b}) = \mathbf{a}, \quad r(\mathbf{a}) = -\mathbf{a}+\mathbf{b} \quad \text{and} \quad r(\mathbf{b}) = -\mathbf{a}.$$

From these, we may find relations between the translations $T_2 = \langle t_a, t_b \rangle$ and the symmetries r and q. They are

$$qt_a = t_b q, \quad qt_b = t_a q, \quad rt_a = t_a^{-1} t_b r \quad \text{and} \quad rt_b = t_a^{-1} r.$$

We know the relations within the groups T_2 and D_3, so the group G of all symmetries is fully specified:

$$G = \{xy : x \in T_2, y \in D_3; \; rt_a = t_a^{-1} t_b r, \; rt_b = t_a^{-1} r, \; qt_a = t_b q, \; qt_b = t_a q\}.$$

Exercise 5.3

Find the point group of the wallpaper group $p3m1$.

Exercise 5.4

Make a copy of the basic parallelogram and its design, shown in Figure 5.6. Mark all the reflection and glide axes and also the centres of rotations of order 3. Write down a representative for each orbit of 3-centres.

5.4 Pattern type $p31m$

There is another type of wallpaper pattern that has rotations of order 3 and reflections of rhombic type in three directions. It has the name $p31m$. From the symbol 1 in the third place, there are no reflections in axes perpendicular to the reference direction, and hence (by Theorem 2.7) no glide reflections in these directions either. The symbol m in the fourth place indicates the presence of reflections in lines parallel to the reference direction. Taking r to be some rotation of order 3, we may form the conjugates by r and r^2 of such a reflection and obtain reflections with axes in other directions. These axes will be parallel to the sides or the short diagonal of a basic parallelogram. In Figure 5.7, we show an example of a pattern of type $p31m$.

Figure 5.7

As usual, we take a pattern of the same type but having a simpler design. In Figure 5.8, we show part of such a pattern together with a basic parallelogram and a generating region.

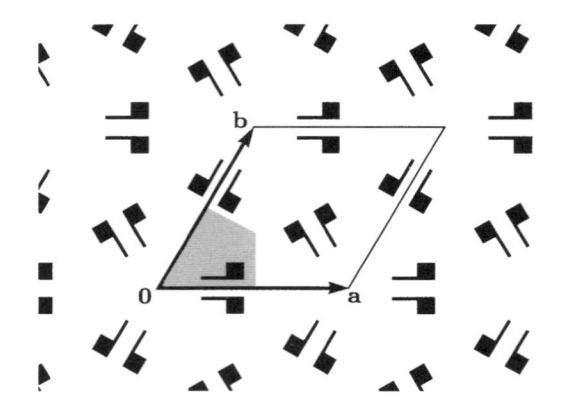

Figure 5.8

As in the case of a pattern of type $p3m1$, we choose our origin O to be the centre of a rotation of order 3 and take $\mathbf{0}$, $\mathbf{a}$, $\mathbf{a}+\mathbf{b}$ and $\mathbf{b}$ to be the vertices of a basic parallelogram. The line through $\mathbf{0}$ and $\mathbf{a}$ will, in this case, be an axis of reflection. We shall let q be this reflection and put $r = r[2\pi/3]$. Then

$$q(\mathbf{a}) = \mathbf{a}, \quad q(\mathbf{b}) = \mathbf{a} - \mathbf{b}, \quad r(\mathbf{a}) = -\mathbf{a} + \mathbf{b} \quad \text{and} \quad r(\mathbf{b}) = -\mathbf{a}.$$

The relations between $T_2 = \langle t_a, t_b \rangle$ and the symmetries r and q will thus be

$$qt_a = t_a q, \quad qt_b = t_a t_b^{-1} q, \quad rt_a = t_a^{-1} t_b r \quad \text{and} \quad rt_b = t_a^{-1} r.$$

The group $\langle r, q \rangle$ will again be the dihedral group D_3 of order 6. Every symmetry of the pattern can be expressed in the form xy, where x is in T_2 and y is in the group $\langle r, q \rangle$.

Thus the group is:

$$G = \{xy : x \in T_2,\, y \in D_3;\ rt_a = t_a^{-1} t_b r,\ rt_b = t_a^{-1} r,\ qt_a = t_a q,\ qt_b = t_a t_b^{-1} q\}.$$

Exercise 5.5 ─────────────────────────────────

Find the point group of the wallpaper group $p31m$.

Exercise 5.6 ─────────────────────────────────

Make a copy of the basic parallelogram and its design, shown in Figure 5.8. Mark all the reflection and glide axes and also the centres of rotations of order 3. Write down a representative for each orbit of 3-centres.

5.5 Pattern type $p3$

The final type of pattern has the name $p3$. A pattern of this type has rotations of order 3 but none of order 6. There are also no reflections or glide reflections. In Figure 5.9, we give an example.

Figure 5.9

The simpler design in Figure 5.10 shows a basic parallelogram and a generating region.

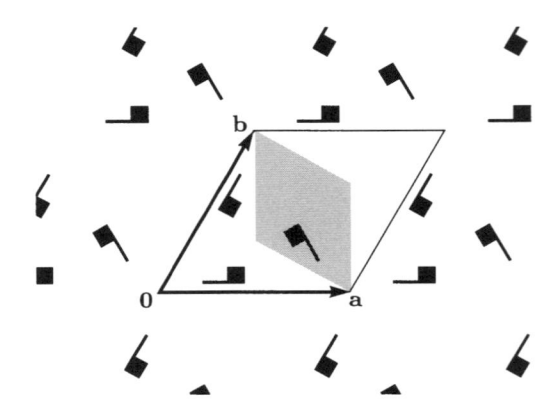

Figure 5.10

Its group of symmetries will consist entirely of direct symmetries, and we may write it as

$$G = \{xy : x \in T_2,\ y \in C_3;\ rt_a = t_a^{-1}t_b r,\ rt_b = t_a^{-1}r\}.$$

Exercise 5.7

Write down a representative for each orbit of 3-centres of the pattern in Figure 5.10.

Exercise 5.8

Find the point group of the wallpaper group $p3$.

We have now investigated all seventeen types of wallpaper patterns. In the final section, we use an audio tape to discuss how we may identify the type of a given pattern. We then collect together many of the properties of the various types of wallpaper pattern.

6 AN ALGORITHM (AUDIO-TAPE SECTION)

This section presents an algorithm, which can be used to classify any wallpaper pattern into one of the seventeen types. In addition to the tape itself, you will need the overlays in the *Geometry Envelope* marked '*Unit GE4* Tape Frame Overlays'.

You should now listen to the audio programme for this unit, referring to the tape frames below when asked to do so during the programme.

1

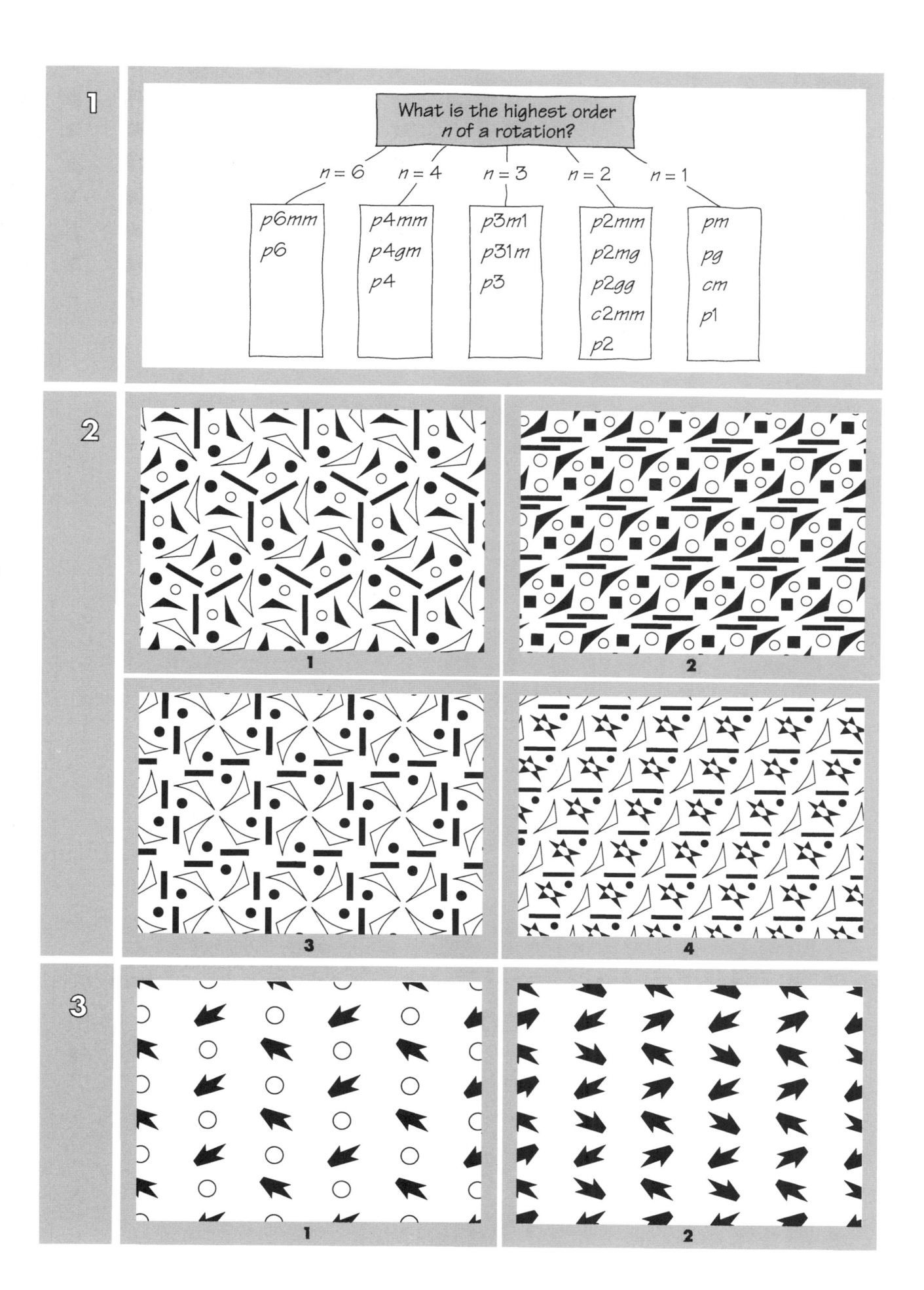

What is the highest order n of a rotation?

- $n = 6$: p6mm, p6
- $n = 4$: p4mm, p4gm, p4
- $n = 3$: p3m1, p31m, p3
- $n = 2$: p2mm, p2mg, p2gg, c2mm, p2
- $n = 1$: pm, pg, cm, p1

3A

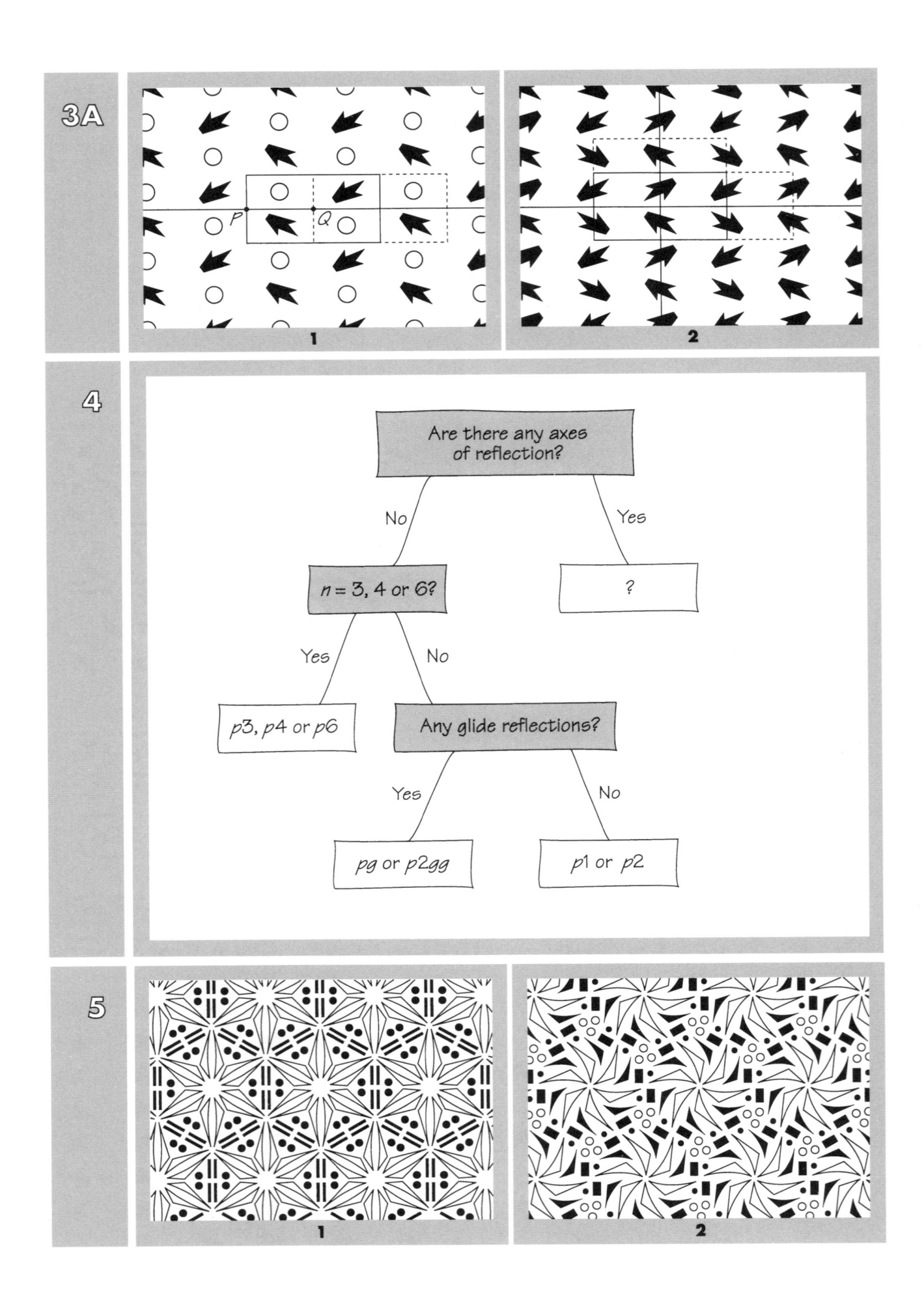

1

2

4

Are there any axes of reflection?

- No → $n = 3, 4$ or 6?
  - Yes → $p3, p4$ or $p6$
  - No → Any glide reflections?
    - Yes → pg or $p2gg$
    - No → $p1$ or $p2$
- Yes → ?

5

1

2

6

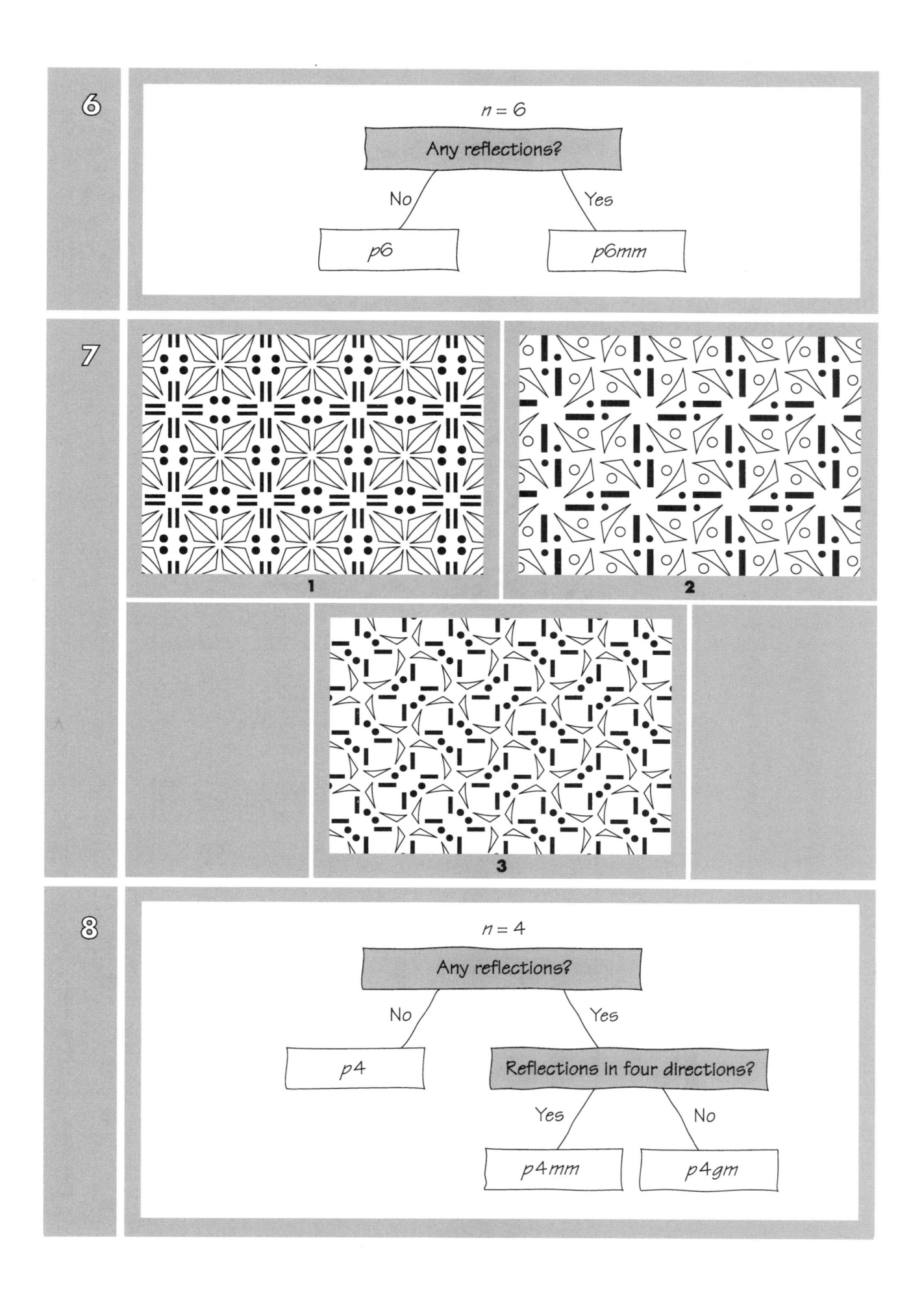

$n = 6$

Any reflections?

- No → *p6*
- Yes → *p6mm*

7

1

2

3

8

$n = 4$

Any reflections?

- No → *p4*
- Yes → Reflections in four directions?
  - Yes → *p4mm*
  - No → *p4gm*

9

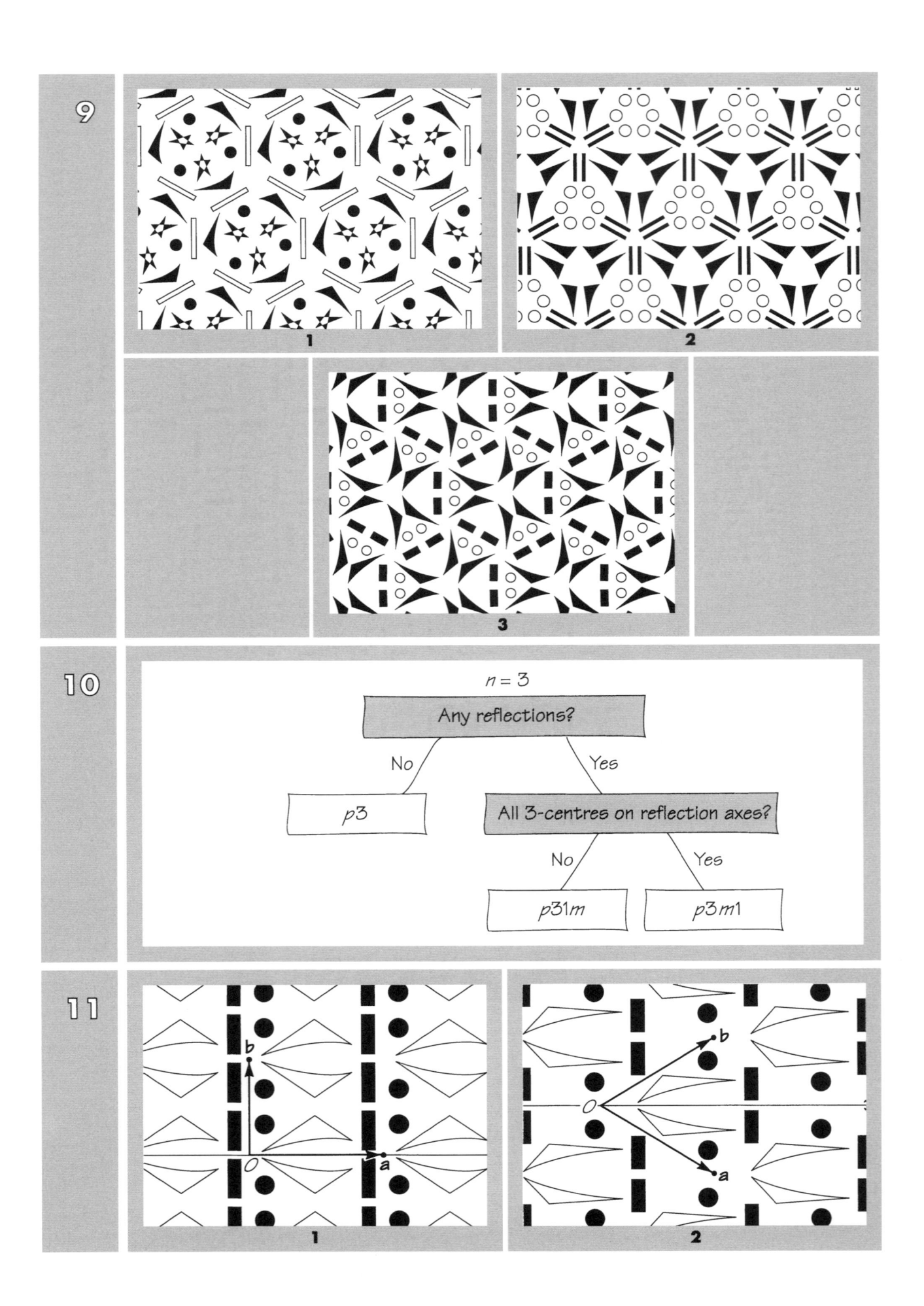

1

2

3

10

$n = 3$

Any reflections?

No — p3

Yes — All 3-centres on reflection axes?

No — p31m

Yes — p3m1

11

1

2

12

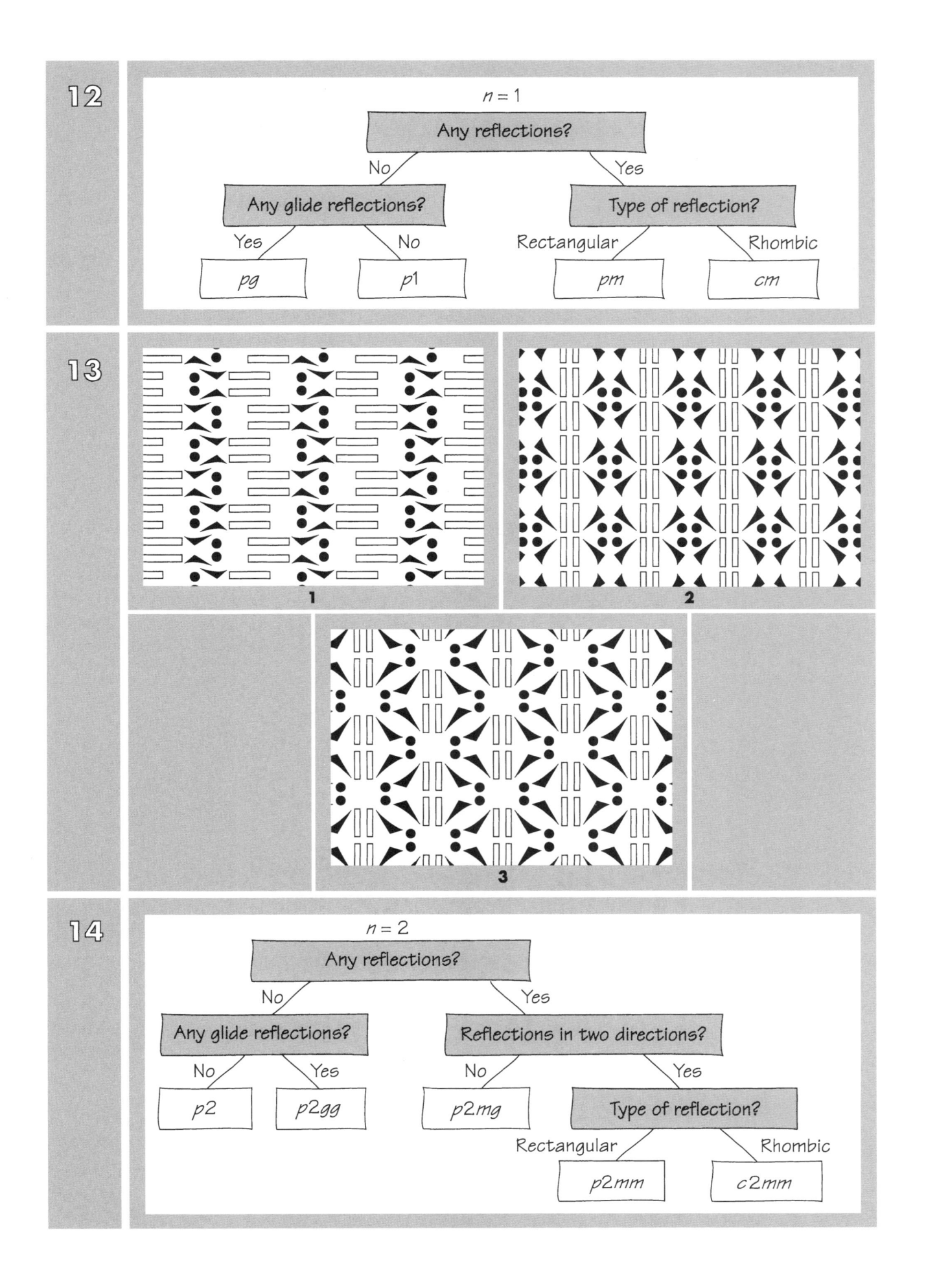

$n = 1$

Any reflections?
- No → Any glide reflections?
  - Yes → **pg**
  - No → **p1**
- Yes → Type of reflection?
  - Rectangular → **pm**
  - Rhombic → **cm**

13

1

2

3

14

$n = 2$

Any reflections?
- No → Any glide reflections?
  - No → **p2**
  - Yes → **p2gg**
- Yes → Reflections in two directions?
  - No → **p2mg**
  - Yes → Type of reflection?
    - Rectangular → **p2mm**
    - Rhombic → **c2mm**

15

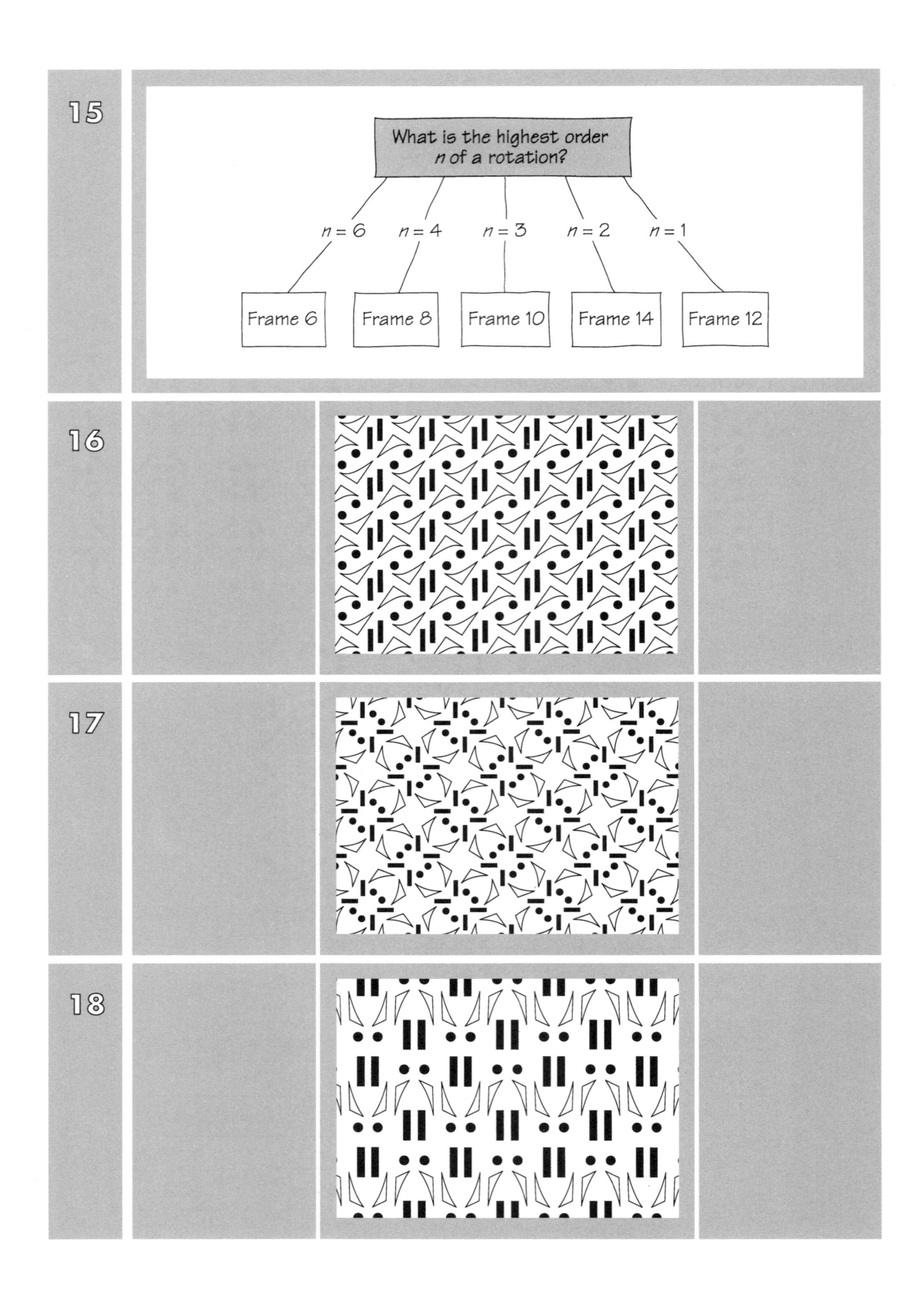

What is the highest order *n* of a rotation?

- *n* = 6 — Frame 6
- *n* = 4 — Frame 8
- *n* = 3 — Frame 10
- *n* = 2 — Frame 14
- *n* = 1 — Frame 12

16

17

18

19

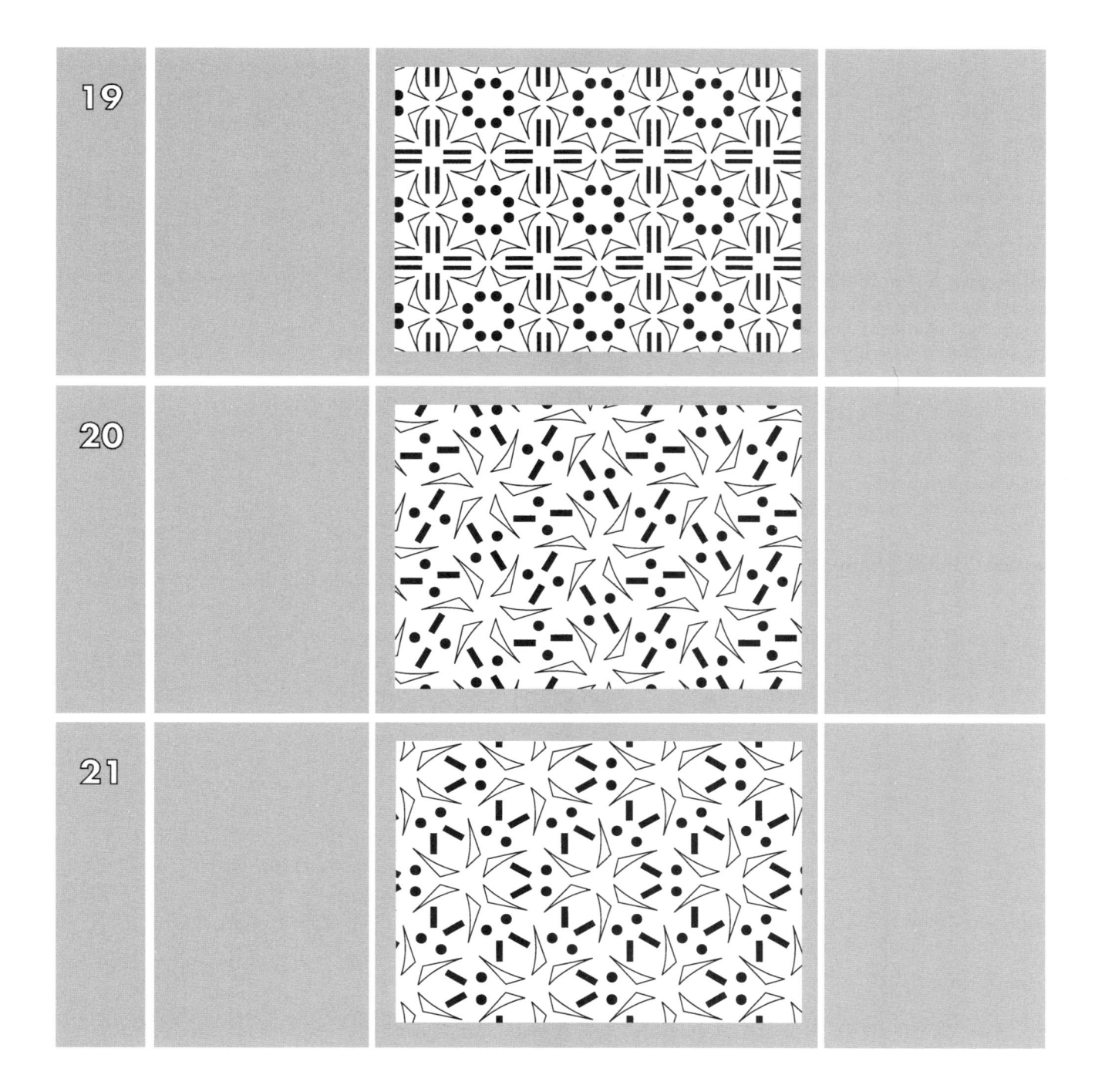

20

21

At the end of the tape, Judith Daniels asks you to use the algorithm, which she describes, to classify the wallpaper patterns given in Frames 16 to 21. The answers are given at the end of the unit, as Solutions 6.1 to 6.6, respectively.

Finally, we promised in Subsection 1.2 to relate our notation for the wallpaper groups to the more abbreviated notation used in the International Tables for X-ray crystallography.

The difference is simply that $p2mm$, $p2mg$ and $p2gg$ are abbreviated to pmm, pmg and pgg respectively (because the existence of axes in two directions automatically implies rotations of order at least 2), and $p4mm$, $p4gm$ and $p6mm$ are abbreviated to $p4m$, $p4g$ and $p6m$ (because in these cases the existence of *any* indirect symmetries forces the existence of certain reflections, so the letter m in the fourth place is redundant).

We now summarize what we have discovered about the wallpaper groups: in the following table, we record our notation (and also the X-ray crystallography notation where it differs), along with the point group and a description of the symmetry group.

Notation	Point group	Symmetry group
$p1$	C_1	$G = T_2$
pm	D_1	$G = \{xy : x \in T_2,\ y \in D_1;\ qt_a = t_a q,\ qt_b = t_b^{-1} q\}$
pg	D_1	$G = \{x, xg : x \in T_2;\ g^2 = t_a,\ gt_a = t_a g,\ gt_b = t_b^{-1} g\}$
cm	D_1	$G = \{xy : x \in T_2,\ y \in D_1;\ qt_a = t_b q,\ qt_b = t_a q\}$
$p2$	C_2	$G = \{xy : x \in T_2,\ y \in C_2;\ rt_a = t_a^{-1} r,\ rt_b = t_b^{-1} r\}$
$p2mm$ (or pmm)	D_2	$G = \{xy : x \in T_2,\ y \in D_2;\ rt_a = t_a^{-1} r,\ rt_b = t_b^{-1} r,\ qt_a = t_a q,\ qt_b = t_b^{-1} q\}$
$p2mg$ (or pmg)	D_2	$G = \{x, xr, xq, xrq : x \in T_2;\ r^2 = e,\ q^2 = e,\ (rq)^2 = t_b,$ $rt_a = t_a^{-1} r,\ rt_b = t_b^{-1} r,\ qt_a = t_a q,\ qt_b = t_b^{-1} q\}$
$p2gg$ (or pgg)	D_2	$G = \{x, xr, xg, xrg : x \in T_2;\ r^2 = e,\ g^2 = t_a,\ (rg)^2 = t_b,$ $rt_a = t_a^{-1} r,\ rt_b = t_b^{-1} r,\ gt_a = t_a g,\ gt_b = t_b^{-1} g\}$
$c2mm$ (or cmm)	D_2	$G = \{xy : x \in T_2,\ y \in D_2;\ rt_a = t_a^{-1} r,\ rt_b = t_b^{-1} r,\ qt_a = t_b q,\ qt_b = t_a q\}$
$p3$	C_3	$G = \{xy : x \in T_2,\ y \in C_3;\ rt_a = t_a^{-1} t_b r,\ rt_b = t_a^{-1} r\}$
$p3m1$	D_3	$G = \{xy : x \in T_2,\ y \in D_3;\ rt_a = t_a^{-1} t_b r,\ rt_b = t_a^{-1} r,\ qt_a = t_b q,\ qt_b = t_a q\}$
$p31m$	D_3	$G = \{xy : x \in T_2,\ y \in D_3;\ rt_a = t_a^{-1} t_b r,\ rt_b = t_a^{-1} r,\ qt_a = t_a q,\ qt_b = t_a t_b^{-1} q\}$
$p4$	C_4	$G = \{xy : x \in T_2,\ y \in C_4;\ rt_a = t_b r,\ rt_b = t_a^{-1} r\}$
$p4mm$ (or $p4m$)	D_4	$G = \{xy : x \in T_2,\ y \in D_4;\ rt_a = t_b r,\ rt_b = t_a^{-1} r,\ qt_a = t_a q,\ qt_b = t_b^{-1} q\}$
$p4gm$ (or $p4g$)	D_4	$G = \{xy, xyg : x \in T_2,\ y \in C_4;\ rt_a = t_b r,\ rt_b = t_a^{-1} r,\ g^2 = t_a,$ $gt_a = t_a g,\ gt_b = t_b^{-1} g,\ (rg)^2 = e\}$
$p6$	C_6	$G = \{xy : x \in T_2,\ y \in C_6;\ rt_a = t_b r,\ rt_b = t_a^{-1} t_b r\}$
$p6mm$ (or $p6m$)	D_6	$G = \{xy : x \in T_2,\ y \in D_6;\ rt_a = t_b r,\ rt_b = t_a^{-1} t_b r,\ qt_a = t_a q,\ qt_b = t_a t_b^{-1} q\}$

SOLUTIONS TO THE EXERCISES

Solution 1.1

One example is the glide reflection whose axis is the vertical axis bisecting the basic rectangle, and whose translation component is translation upwards by half the height of the basic rectangle.

In other words, with the coordinate system shown in the figure, the glide reflection

$$q[(0,1),(0,0),\pi/2]$$

is a symmetry of W but not a symmetry of any associated lattice.

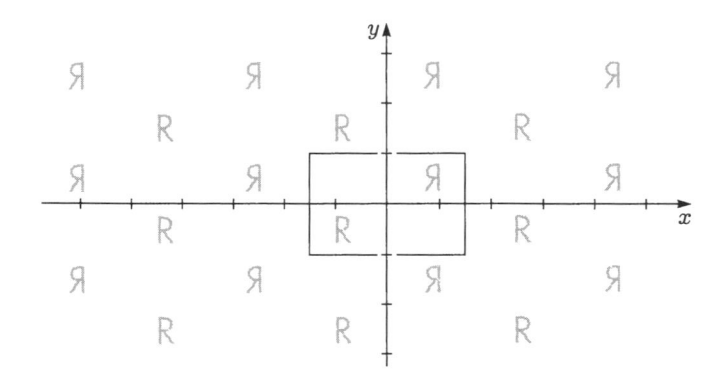

Solution 1.2

There are many possible choices of a generating region (infinitely many, in fact!). We show two examples below.

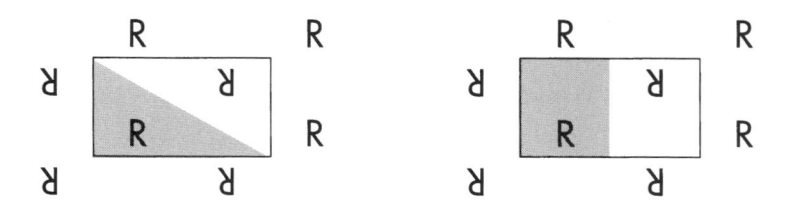

Solution 1.3

An example of a generating region is shown below. Notice that any non-trivial symmetry of the pattern, applied to this generating region, will take some of it outside the marked basic rectangle.

Solution 1.4

There are six directions of reflection axes, namely those making angles with the x-axis that are multiples of $\pi/6$. The first letter m corresponds to those that are perpendicular to the reference direction. Now the reference direction makes an angle of $-2\pi/3$ with the x-axis, so the direction perpendicular to this makes an angle of $-2\pi/3 + \pi/2 = -\pi/6$ (or, equivalently, $5\pi/6$) with the x-axis. The existence of rotations of order 6 means that we can add multiples of $\pi/3$ to this.

Therefore, the first m refers to reflection axes at angles of $\pi/6$, $\pi/2$ and $5\pi/6$ (or $-\pi/6$) with the x-axis, and the second m refers to reflection axes in the direction of the x-axis itself, and at angles of $\pi/3$ and $2\pi/3$ with the x-axis.

Solution 1.5

The notation is $p1$.

Solution 2.1

A wallpaper patter W with symmetry group G clearly has the same type as itself, since the identity map e is affine and the mapping $g \mapsto ege^{-1}$ is the identity map from G to itself. The relation is therefore reflexive.

Suppose that W' has the same type as W where the symmetry groups are G' and G respectively. Then there is an affine transformation ϕ such that the mapping $g \mapsto g' = \phi g \phi^{-1}$ is a mapping of G onto G'. The inverse map ϕ^{-1} will also be affine, and will satisfy $g = (\phi^{-1})g'(\phi^{-1})^{-1}$, and this gives us a mapping from G' onto G. Thus the relation is symmetric.

Finally, suppose we have three wallpaper patterns W, W' and W'' with symmetry groups G, G' and G'', respectively. If W' has the same type as W and W'' has the same type as W', there will be affine maps ϕ and ψ such that $g \mapsto g' = \phi g \phi^{-1}$ is from G onto G' and $g' \mapsto g'' = \psi g' \psi^{-1}$ is from G' onto G''. The composition $\psi\phi$ will be affine, and it will give us the mapping $g \mapsto g'' = (\psi\phi)g(\psi\phi)^{-1}$ from G onto G''. Thus the relation is transitive.

As the relation is reflexive, symmetric and transitive, it is an equivalence relation.

Solution 2.2

W has no rotational symmetries of order 4, and no diagonal reflection symmetries. It does have rotational symmetries of order 2, and horizontal and vertical reflection symmetries. Therefore, there is an affine transformation ϕ mapping W to a pattern W' of the same type but with a rectangular associated lattice.

Solution 2.3

The symmetry f is given by

$$\begin{aligned} f &= r[(1,1), \pi/2] \\ &= t[(1,1) - r[\pi/2](1,1)] \, r[\pi/2] \quad \text{(by Equation 8 of the Isometry Toolkit)} \\ &= t[(1,1) - (-1,1)] \, r[\pi/2] \\ &= t[(2,0)] \, r[\pi/2]. \end{aligned}$$

Thus the translation part of f is $t[(2,0)]$, which is not in the group $\Delta(W) = \langle t[(4,0)], t[(0,4)] \rangle$.

Solution 2.4

The point group is $C_2 = \{e, r[\pi]\}$.

Solution 2.5

The point group is D_1, consisting of the identity and a reflection. (Using a coordinate system whose x-axis is horizontal, the point group is $\{e, q[\pi/2]\}$.)

Note that the reflection is not a symmetry of the pattern.

Solution 2.6

The elements of G^+/T_2 are the cosets of T_2 in G^+. In *Unit IB4* these were normally written as left cosets, so one would have

$$G^+/T_2 = \{T_2, rT_2, r^2 T_2, \ldots, r^{n-1} T_2\}$$

where n is the order of r. However, right and left cosets are the same for normal subgroups, so equally good would be

$$G^+/T_2 = \{T_2, T_2 r, T_2 r^2, \ldots, T_2 r^{n-1}\}.$$

Solution 2.7

(a) The axis of f is the line $y = 1$. The translation component is therefore $t[(4,0)]$, which is a symmetry of W. Therefore the reflection component is also a symmetry of W, and the axis is a reflection axis of W.

(b) The axis of g is the line $x = 1$. The translation component is therefore $t[(0,0)]$. That is to say, g is actually a reflection — so the axis is certainly a reflection axis.

(c) Some more explicit calculation is called for here. We need to express h as

$$\begin{aligned} h &= q[\mathbf{g}, \mathbf{c}, \pi/4] \\ &= t[\mathbf{g} + 2\mathbf{c}] \, q[\pi/4] \end{aligned}$$

where $\mathbf{c}$ is perpendicular to the reflection axis (see Equation 15 of the Isometry Toolkit). Since $\mathbf{g} + 2\mathbf{c} = (0,4)$, we have

$$\mathbf{g} = (2,2), \quad 2\mathbf{c} = (-2,2),$$

so $\mathbf{c} = (-1,1)$, and the axis passes through $(-1,1)$ and is inclined at $\pi/4$ to the x-axis.

The translation component is $t[(2,2)]$, which is not a symmetry of W. Therefore, the reflection component is not a symmetry either, and the axis is a glide axis.

Solution 2.8

The indirect symmetries with axes in the x- and y-directions are rectangular, and those with axes in diagonal directions are rhombic. Therefore f and g are rectangular, and h is rhombic.

Solution 2.9

Let q be the linear part of f, q' the linear part of g. Since the axes are at right angles, we can express q' as

$$q' = r[\pi] \, q.$$

First suppose that f is rectangular. Then there is a basis $\{\mathbf{a}, \mathbf{b}\}$ of the associated lattice L based at O, such that

$$q(\mathbf{a}) = \mathbf{a}, \quad q(\mathbf{b}) = -\mathbf{b}.$$

Then

$$q'(\mathbf{a}) = r[\pi] \, q(\mathbf{a})$$
$$= r[\pi](\mathbf{a})$$
$$= -\mathbf{a},$$
$$q'(\mathbf{b}) = r[\pi] \, q(\mathbf{b})$$
$$= r[\pi](-\mathbf{b})$$
$$= \mathbf{b}.$$

Therefore, by letting $\mathbf{a}' = \mathbf{b}$ and $\mathbf{b}' = \mathbf{a}$, we see that g is rectangular.

Next, suppose that f is rhombic. Then there is a basis $\{\mathbf{a}, \mathbf{b}\}$ of L such that

$$q(\mathbf{a}) = \mathbf{b}, \quad q(\mathbf{b}) = \mathbf{a}.$$

Then

$$q'(\mathbf{a}) = r[\pi] \, q(\mathbf{a})$$
$$= r[\pi](\mathbf{b})$$
$$= -\mathbf{b},$$
$$q'(\mathbf{b}) = r[\pi] \, q(\mathbf{b})$$
$$= r[\pi](\mathbf{a})$$
$$= -\mathbf{a}.$$

Thus $q'(-\mathbf{b}) = \mathbf{a}$. Therefore, by letting $\mathbf{a}' = \mathbf{a}$ and $\mathbf{b}' = -\mathbf{b}$, we see that g is rhombic.

Solution 3.1

The point group is C_2.

Solution 3.2

The point group is $\{e\} = C_1$.

Solution 3.3

The symmetries that map the given generating region into the basic rectangle are e, $t_b q$, $t_a r q$ and $t_a t_b r$.

Solution 3.4

The point group is D_2.

Solution 3.5

We may take the points $\mathbf{0}$ and $\frac{1}{2}\mathbf{a}$ as representatives of the two orbits of 2-centres. Then $\text{Stab}(\mathbf{0}) = \{e, r\}$ and $\text{Stab}(\frac{1}{2}\mathbf{a}) = \{e, t_a r\}$.

Solution 3.6

Since $g'^2 = t_b = t[\mathbf{b}]$, the translation component of g' is $t[\frac{1}{2}\mathbf{b}]$.

Solution 3.7

The point group consists of the identity e, the rotation r and the reflections $q = t[\frac{1}{2}(\mathbf{b} - \mathbf{a})] \, g$ and rq. It is the group D_2.

Solution 3.8

The point group consists of e and q; it is D_1.

Solution 3.9

The point group consists of e and the linear part of g, namely q. It is D_1.

Solution 4.1

The point group is D_2.

Solution 4.2

The point group consists of e and q; it is D_1.

Solution 4.3

The point group is D_4.

Solution 4.4

(a)

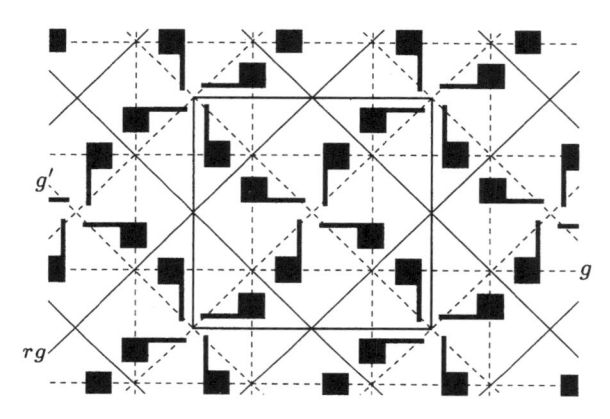

Reflection axes are shown as solid lines and glide axes as broken lines.

The vector $\frac{1}{2}\mathbf{a}$ is parallel to the axis of q and $\frac{1}{2}\mathbf{b}$ is perpendicular to it, so the translation component of g is $t[\frac{1}{2}\mathbf{a}]$.
$-\frac{1}{2}\mathbf{a} + \frac{1}{2}\mathbf{b}$ is parallel to the axis of q', so the translation component of g' is $t[-\frac{1}{2}\mathbf{a} + \frac{1}{2}\mathbf{b}]$.

(b) We have $g = t[\frac{1}{2}\mathbf{a} + \frac{1}{2}\mathbf{b}]\,q$, $\quad g' = t[-\frac{1}{2}\mathbf{a} + \frac{1}{2}\mathbf{b}]\,q'\quad$ and $\quad q' = qr$.

Calculating, we get

$$gg' = t[\tfrac{1}{2}\mathbf{a} + \tfrac{1}{2}\mathbf{b}]\,q\,t[-\tfrac{1}{2}\mathbf{a} + \tfrac{1}{2}\mathbf{b}]\,qr$$
$$= t[\tfrac{1}{2}\mathbf{a} + \tfrac{1}{2}\mathbf{b}]\,t[-\tfrac{1}{2}\mathbf{a} - \tfrac{1}{2}\mathbf{b}]\,qqr$$
$$= r.$$

Solution 4.5

$$g^2 = t[\tfrac{1}{2}\mathbf{a} + \tfrac{1}{2}\mathbf{b}]\, q\, t[\tfrac{1}{2}\mathbf{a} + \tfrac{1}{2}\mathbf{b}]\, q$$
$$= t[\tfrac{1}{2}\mathbf{a} + \tfrac{1}{2}\mathbf{b}]\, t[\tfrac{1}{2}\mathbf{a} - \tfrac{1}{2}\mathbf{b}]\, q^2$$
$$= t[\mathbf{a}].$$
$$gt[\mathbf{a}] = t[\tfrac{1}{2}\mathbf{a} + \tfrac{1}{2}\mathbf{b}]\, q\, t[\mathbf{a}]$$
$$= t[\tfrac{1}{2}\mathbf{a} + \tfrac{1}{2}\mathbf{b}]\, t[\mathbf{a}]\, q$$
$$= t[\mathbf{a}]\, (t[\tfrac{1}{2}\mathbf{a} + \tfrac{1}{2}\mathbf{b}]\, q)$$
$$= t[\mathbf{a}]\, g.$$
$$gt[\mathbf{b}] = t[\tfrac{1}{2}\mathbf{a} + \tfrac{1}{2}\mathbf{b}]\, g\, t[\mathbf{b}]$$
$$= t[\tfrac{1}{2}\mathbf{a} + \tfrac{1}{2}\mathbf{b}]\, t[-\mathbf{b}]\, q$$
$$= t[-\mathbf{b}]\, (t[\tfrac{1}{2}\mathbf{a} + \tfrac{1}{2}\mathbf{b}]\, q)$$
$$= (t[\mathbf{b}])^{-1} g.$$
$$(rg)^2 = r\, t[\tfrac{1}{2}\mathbf{a} + \tfrac{1}{2}\mathbf{b}]\, q\, r\, t[\tfrac{1}{2}\mathbf{a} + \tfrac{1}{2}\mathbf{b}]\, q$$
$$= t[\tfrac{1}{2}\mathbf{b} - \tfrac{1}{2}\mathbf{a}]\, r\, q\, t[\tfrac{1}{2}\mathbf{b} - \tfrac{1}{2}\mathbf{a}]\, r q$$
$$= t[\tfrac{1}{2}\mathbf{b} - \tfrac{1}{2}\mathbf{a}]\, r\, t[-\tfrac{1}{2}\mathbf{b} - \tfrac{1}{2}\mathbf{a}]\, qrq$$
$$= t[\tfrac{1}{2}\mathbf{b} - \tfrac{1}{2}\mathbf{a}]\, t[\tfrac{1}{2}\mathbf{a} - \tfrac{1}{2}\mathbf{b}]\, (rq)^2$$
$$= e.$$

Solution 4.6

The linear parts of the elements of G form the group D_4, which is the point group.

Solution 4.7

The point group is C_4.

Solution 5.1

The point group is D_6.

Solution 5.2

The point group is C_6.

Solution 5.3

The point group is D_3.

Solution 5.4

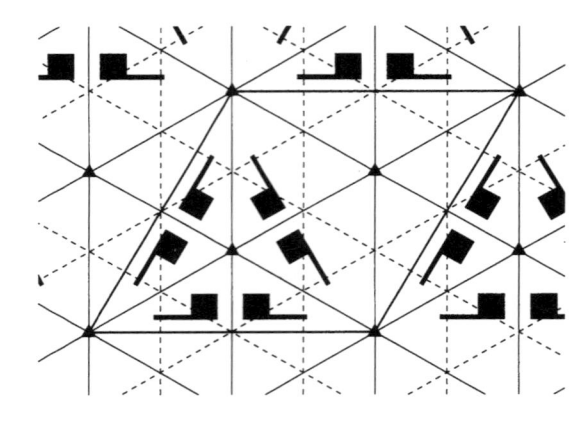

Reflection axes are shown as solid lines and glide axes as broken lines.

The 3-centres are shown as small triangles in the figure. There are three orbits of 3-centres. We may take as representatives the points $\mathbf{0}$, $\tfrac{1}{3}\mathbf{a} + \tfrac{1}{3}\mathbf{b}$ and $\tfrac{2}{3}\mathbf{a} + \tfrac{2}{3}\mathbf{b}$. Notice that all of the 3-centres lie on reflection axes of the pattern.

Solution 5.5

The point group is D_3.

Solution 5.6

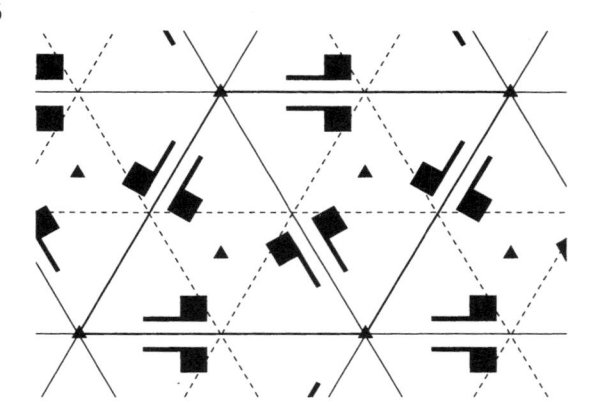

Figure S.1

The 3-centres are shown as small triangles in the figure. There are two orbits of 3-centres. We may take $\mathbf{0}$ and $\frac{1}{3}\mathbf{a} + \frac{1}{3}\mathbf{b}$ as representatives. (In this case, the point $\frac{2}{3}\mathbf{a} + \frac{2}{3}\mathbf{b}$ lies in the same orbit as the point $\frac{1}{3}\mathbf{a} + \frac{1}{3}\mathbf{b}$.)

Notice that the 3-centres in the orbit represented by $\mathbf{0}$ lie on reflection axes, while those in the other orbit do not.

Solution 5.7

There are three orbits of 3-centres represented by the points $\mathbf{0}, \frac{1}{3}\mathbf{a} + \frac{1}{3}\mathbf{b}$ and $\frac{2}{3}\mathbf{a} + \frac{2}{3}\mathbf{b}$.

Solution 5.8

The point group is C_3.

Solution 6.1

The highest order of a rotation is 2. (For example, the centre of the basic parallelogram on Overlay 16 is such a centre.) Thus we go to Frame 14.

There are no reflections, and no glide reflections. Therefore the type of pattern in Frame 16 is $p2$.

Solution 6.2

The highest order of a rotation is 4. (Again, the centre of the basic parallelogram on Overlay 17 is such a centre.) Thus we go to Frame 8.

There are reflections, but not in four directions. (The reflection axes are diagonal to the page; typical reflection axes bisect the sides of the basic parallelogram on the overlay.) Therefore the type of pattern in Frame 17 is $p4gm$.

Solution 6.3

The highest order of a rotation is 2. (For example, the points $\frac{1}{4}$ and $\frac{3}{4}$ of the way along a horizontal side of the basic parallelogram on Overlay 18 are such centres.) Thus we go to Frame 14.

There are reflections. (For example, the vertical sides of the basic parallelogram on the overlay are reflection axes.) However, they are not in two directions. Therefore the type of pattern in Frame 18 is $p2mg$.

Solution 6.4

The highest order of a rotation is 4. (For example, the centre of the basic parallelogram on Overlay 19 is such a centre.) Thus we go to Frame 8.

There are reflections in four directions. Therefore the type of pattern in Frame 19 is *p4mm*.

Solution 6.5

The highest order of a rotation is 6. (For example, the vertices of the basic parallelogram on Overlay 20 are such centres.) Thus we go to Frame 6.

There are no reflections. (The triangle-like shapes in the pattern have no reflection symmetries, and they are all rotated images of each other.) Therefore the type of pattern in Frame 20 is *p6*.

Solution 6.6

The highest order of a rotation is 3. (For example, the vertices of the basic parallelogram on Overlay 21 are such centres.) Thus we go to Frame 10.

There are reflections. (For example, the short diagonal of the basic parallelogram on the overlay lies on a reflection axis.) However, not all 3-centres lie on reflection axes. (For example, the points $\frac{1}{4}$ and $\frac{3}{4}$ of the way along the long diagonal of the basic parallelogram are 3-centres that do not lie on reflection axes.) Therefore the type of pattern in Frame 21 is *p31m*.

OBJECTIVES

After you have studied this unit, you should be able to:

(a) understand the basic terminology used, such as wallpaper pattern, associated lattice, basic parallelogram, design, generating region and point group;

(b) recognize and be thoroughly familiar with the different types of symmetries that a wallpaper pattern may possess, and know the distinction between indirect symmetries of rectangular and rhombic type;

(c) decide whether or not two given wallpaper patterns are of the same type;

(d) use the algorithm to identify the type of a given wallpaper pattern.

INDEX

associated lattice 6
associated lattice based at O 6
axes
  rectangular 22
  rhombic 22
basic parallelogram 7
design 7

generating region 8
lattice, associated 6
point group 18
rectangular axes 22
rectangular symmetry 21
rhombic axes 22
rhombic symmetry 21

seventeen wallpaper groups 23
symmetry
  rectangular 21
  rhombic 21
type of wallpaper pattern 14
wallpaper group 10
wallpaper pattern 6